Liang An
Editor

Recycling of Spent Lithium-Ion Batteries

Processing Methods and Environmental Impacts

Springer

Editor
Liang An
Department of Mechanical Engineering
The Hong Kong Polytechnic University
Hong Kong, China

ISBN 978-3-030-31833-8 ISBN 978-3-030-31834-5 (eBook)
https://doi.org/10.1007/978-3-030-31834-5

This Springer imprint is published by the registered company Springer Nature Switzerland AG
The registered company address is: Gewerbestrasse 11, 6330 Cham, Switzerland

Preface

Renewable energy is the final solution to mitigate climate change resulted from the burning of fossil fuels that produces large quantities of greenhouse gases. Currently, lithium-ion batteries (LIBs) are the most widely used batteries in portable devices, electric vehicles, and grid energy storage due to their high-energy and high-power densities, simple operation, and relatively long lifetime. Owing to the constant offers of upgrades in consumer electronics and rapid development of electric vehicles, gigantic amount of LIBs have been manufactured in the recent decades. Moreover, the ever-increasing need for electric vehicles will further accelerate the production of LIBs. Hence, predictably, we have to face serious disposal problems of considerable amount of spent LIBs in the near future if no appropriate recycling processes can be implemented. The impact of spent LIBs on the environment mainly involves two aspects. On the one hand, most of the cathode materials contain hazardous heavy metals (e.g., Co) and harmful organic electrolytes, which can cause serious environmental pollution. On the other hand, spent LIBs contain high-value metals, such as Li, Co, Ni, Cu, and Al. Some of them, such as Co and Li, are much more abundant in LIBs than in natural ores. Therefore, recycling spent LIBs is highly desirable since it can not only reduce environmental pollution, but also preserve mineral resources. This book offers in-depth coverage of the latest advances in recycling spent LIBs, and the major features are summarized as follows:

Chapter "Recycling Technology and Principle of Spent Lithium-Ion Battery" demonstrates environmental risk of spent LIBs, treatment of spent LIBs, and cascade utilization of spent LIBs. The development history of China's power battery cascade utilization policy is introduced.

Chapter "Hydrometallurgically Recycling Spent Lithium-Ion Batteries" discusses the difference between hydrometallurgical methods and pyrometallurgical methods for recycling spent LIBs. This chapter concentrates on the leaching process, which is a key and indispensable step for the recovering valuable resources from spent LIBs in the whole hydrometallurgical processes. The purification and recovery processes are also presented.

Chapter "Pyrometallurgical Routes for the Recycling of Spent Lithium-Ion Batteries" introduces the mechanisms and equipment of pyrometallurgical routes for the recycling of spent LIBs. The combination of the pyro-methods with hydro-methods is also highlighted, which is a greener and more efficient process.

Chapter "Bio-hydrometallurgically Treatment of Spent Lithium-Ion Batteries" describes the mechanisms and impacts on recovery efficiency of bio-hydrometallurgical methods for recycling spent LIBs. The pros and cons of bio-hydrometallurgical methods are highlighted.

Chapter "Hydrometallurgical Processes for Valuable Metals Recycling from Spent Lithium-Ion Batteries" gives a relatively systematic framework concerning the leaching of metals from spent LIBs for the discussion and evaluation of the current leaching methods, including chemical leaching and bioleaching. The separation and recovery of metals from leaching solutions are comprehensively summarized and discussed concerning chemical precipitation, solvent extraction, and other separation methods. The regeneration of cathode materials by hydrometallurgical method is introduced and analyzed as well.

Chapter "High Value-Added Products From Recycling of Spent Lithium-Ion Batteries" summarizes the high value-added products from recycling of spent LIBs. Some typical recovery routes have been discussed in detail.

Chapter "Bio-hydrometallurgical Methods For Recycling Spent Lithium-Ion Batteries" reviews the information available on the basic principles and recent developments of the bioleaching of metals from spent LIBs in detail. The challenges, limitations, and potential solutions for applying more efficient bioleaching approach for recovery of metals from LIBs are also covered in this chapter.

Chapter "Impacts of Recycling of Spent Lithium-Ion Batteries on Environmental Burdens" gives a systematic overview of this issue, including potential risks of spent LIBs, the assessment methodology and evaluation results of environmental burdens of spent LIBs, the discussion on the environmental impact of recycling process, and the approaches for designing an environmental benign recycling process.

This book is an essential reference resource for professionals, researchers, and policymakers around the globe working in academia, industry, and government. This work was fully supported by a grant from the Research Grants Council of the Hong Kong Special Administrative Region, China (Project No. 15222018). We would like to express our gratitude to all the authors who submitted their contributions and shared valuable state-of-the-art knowledge and experience on associated topics for publication in this book. In addition, we are grateful to all the reviewers who helped to improve the contributions. Furthermore, we also would like to thank Mr. Zhefei Pan for his assistance in preparing this book. Last but not least, we acknowledge the professional staff from Springer for their continuous support.

Hong Kong, China Liang An

Contents

Recycling Technology and Principle of Spent Lithium-Ion Battery

Siqi Zhao, Wenzhi He and Guangming Li

1 Introduction

With the continuous improvement of people's living standards, Electrical and Electronic Equipment (EEE) has become an indispensable part of people's daily life, and the penetration rate has increased rapidly. It is also the increase in the use frequency of household appliances that directly leads to the annual increase in the number of Waste Electric and Electronic Equipment (WEEE). Efficient and environment-friendly disposal of waste electrical and electronic products is not only a rigid demand for environmental protection, but also an important measure to actively cultivate new growth points and promote green development under the new normal with great downward pressure on the economy and increasing constraints on resources and environment [1].

Since the 1990s, lithium-ion batteries have been widely used in the field of portable electronic products due to their high charging voltage, high specific energy, long cycle life, good safety performance, low pollution, no memory effect, and small self-discharge. Including mobile phones, laptops, camcorders, digital cameras, medical devices, etc. [2], and has an important role in the fields of electric vehicles (HEVs, PHEVs, EVs), energy storage batteries for large power plants, UPS power supplies, medical instrument power supplies and even space [3].

In 2016, the global lithium-ion battery market scale exceeded 90 GW h, with a year-on-year growth of 18%. The industrial scale reached at $37.8 billion, with a year-on-year growth of 16% [4]. With the booming development of new energy vehicles, the global lithium-ion battery market will also show explosive growth (Fig. 1). In

S. Zhao · W. He · G. Li (✉)
College of Environmental Science and Engineering, Tongji University, Shanghai 200092, People's Republic of China
e-mail: ligm@tongji.edu.cn

Shanghai Institute of Pollution Control and Ecological Security, Shanghai 200092, People's Republic of China

L. An (ed.), *Recycling of Spent Lithium-Ion Batteries*,
https://doi.org/10.1007/978-3-030-31834-5_1

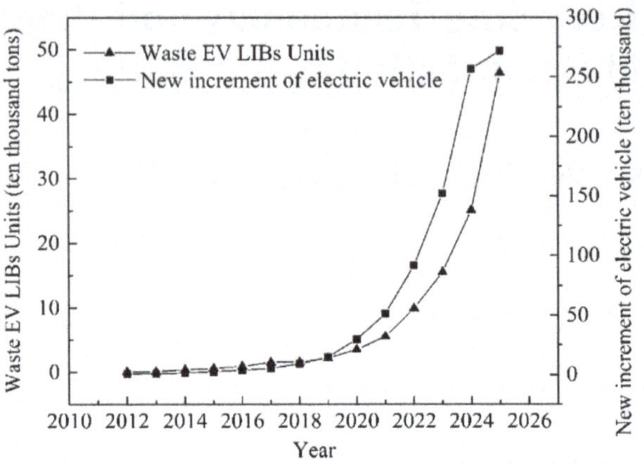

Fig. 1 China's annual production of new incremental electric vehicles and discarded power batteries [8]

2012, the number of new electric vehicles increased by 7700, and in 2015, it increased to 2,717,100. The widespread use of lithium batteries and the gradually shortened replacement time lead to the generation of a large number of discarded lithium batteries [5, 6]. It is predicted that the output of discarded power batteries will increase from 10,700 tons in 2012 to 464,000 tons in 2025, with a compound annual growth rate of 59%. By 2020, the number of spent lithium-ion batteries produced in China will exceed 25 billion, and the weight will exceed 500,000 tons [7].

The lithium-ion battery is mainly composed of shell, anode, cathode, separator and electrolyte and other components. Figure 2 has shown the composition of spent lithium-ion battery.

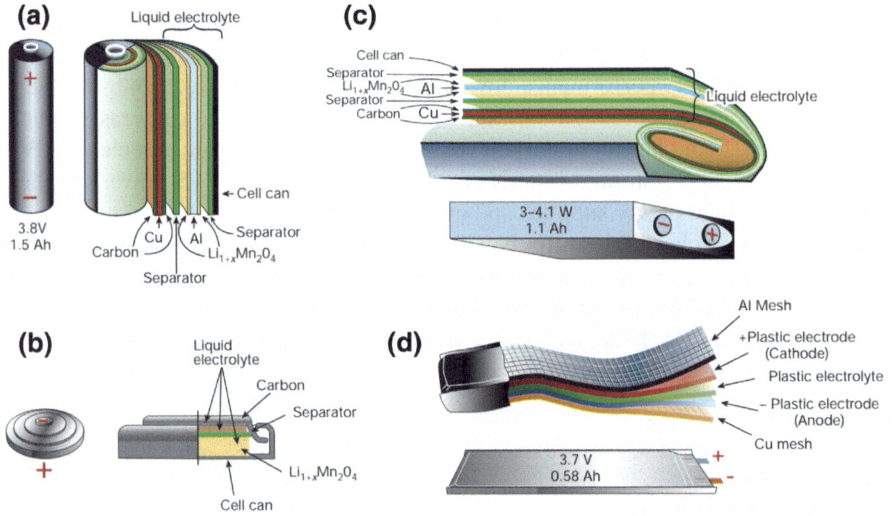

Fig. 2 Composition of spent lithium-ion battery. **a**, Cylindrical; **b**, coin; **c**, prismatic; and **d**, thin and flat [9]

Shell: The shell of lithium battery is usually stainless steel or nickel-plated steel with single component. After mechanical separation, due to its high purity can be directly concentrated recovery, the subsequent resource is more convenient.

Cathode: The cathode active materials of lithium-ion battery are metal oxides containing lithium, such as $LiCoO_2$, $LiNiO_2$, $LiMn_2O_4$, $LiPePO_4$ and $LiNi_xCo_yMn_zO_2$. The rapid development of electric vehicles (EVs), hybrid vehicles (HEVs) and plug-in hybrid vehicles (PHEVs) has prompted researchers to search for high energy density of electrode materials of lithium-ion batteries [10, 11]. Although many new cathode materials with high voltage, they are kind of typical cathode material. Because of their high energy density, easy production and persistence, $LiCoO_2$ has been widely used in mobile electronic devices, such as phones, laptops, watches, digital cameras, communications equipment, etc. [12–14]. The cathode composition of commercial lithium-ion batteries is generally about 90% of $LiCoO_2$, 8% conductive agent and 4% polyvinylidene fluoride (PVDF) binder. The cathode active material is uniformly coated on both sides of the aluminum foil collector by PVDF binder.

Anode: The anodes also have the active materials (such as graphite) coated by PVDF on copper foils collector with a thickness of 15 μm which has the similar structure to that of the cathodes [15–18]. A very thin layer of PVDF has formed in the area of conductive agent (acetylene black) surface and active material particles, so that to ensure enough bond of active materials between particles and acetylene black. At the same time, the free binder between active material particles and acetylene black can provide enough bond energy, firmly stick to the active material particles on the collection of fluid [13]. Carbon anode materials mainly include: graphite, petroleum coke, carbon fiber, pyrolysis carbon, mesophase bituminous carbon microsphere (MCMB), carbon black, glass carbon, etc. In recent years, in order to increase the storage capacity of lithium-ion batteries, some new cathode active materials have been developed, while graphite is still the most commonly used cathode active material in commercial lithium-ion batteries. Therefore, many studies currently focus on modifying the structure of graphite to improve its properties [19–22].

Electrolyte: The lithium-ion battery electrolyte plays the role of transferring charge between the cathode and anode in the battery, and is essential for the specific capacity of the battery, the operating temperature range, the cycle efficiency and the safety performance, mainly consists of high-purity organic solvents and electrolyte lithium salts. There are currently four types of electrolytes for lithium ion batteries, namely liquid electrolytes, colloidal electrolytes, polymer electrolytes, and ceramic electrolytes. The most common liquid electrolyte is a liquid in which a lithium salt is dissolved in an organic solvent, such as carbonate. At present, the most commonly used lithium ion battery electrolyte salts are $LiPF_6$, $LiBF_4$, $LiCF_3SO_3$, $Li(SO_2CF_3)_2$ [16]. Most of the lithium-ion battery electrolytes apply LiPF6 as the salt, with a solution that has high conductivity (>10^3 s/cm), high ion mobility (0.35), and acceptable safety [23, 24]. In commercial batteries, the main organic solvents are propylene carbonate (PC), vinyl carbonate (EC), dimethyl carbonate (DMC), diethyl carbonate (DEC) and methyl ethyl carbonate (EMC) [16, 25]. Currently, most electrolytes used in commercial lithium-ion batteries are mixtures of several carbonates, such as dimethyl carbonate (DMC) or methyl ethyl carbonate (EMC) [16]. At present, most electrolytic liquids used by domestic enterprises in China are a combination of several carbonates, such as EC + DEC (3:7) mixed solvents.

Separator: In order to prevent short circuit, a separator is generally used to separate the cathode and anode. The performance of the separator determines the interface structure and internal resistance of the battery, which directly affects the capacity, cycle and safety performance of the battery. The separator with excellent performance plays an important role in improving the overall performance of the battery. The most widely used separators today are polyolefin-based materials such as polyethylene (PE) and polypropylene (PP) with acceptable cost, proper pore structure, excellent mechanical strength, and good overheat protection properties [26]. In addition, some porous films contain two or more materials and have good mechanical properties up to 165 °C. However, when passing through the polyolefin membrane, the migration process of lithium ions has impaired, which seriously affect the electrochemical energy efficiency of the battery. Therefore, in order to overcome this shortcoming, various surface modification methods have been tried, such as plasma treatment, graft polymerization, gel polymer electrolyte dip coating and infiltration, to improve the thermal stability of commercial polyolefin membranes [27–30].

The composition of lithium batteries produced by different manufacturers may be slightly different, usually containing about 5–20% of Co, 5–10% of Ni, 5–7% of Li, about 15% of organic matter, and about 7% of plastic. It has been estimated that one lithium-ion battery in mobile phone weighing about 40 g, the content of Co, Cu, Al, Fe and Li metals are 15, 14, 4.7, 25 and 0.1% respectively (as shown in Fig. 3) [31]. Organics include electrolytes and membranes [32].

Lithium-ion batteries contain heavy metals, organic electrolytes, and organic electrolytes that are highly toxic. On the one hand, improper disposal of discarded lithium batteries may result in environmental risks of heavy metals and electrolytes, and may have adverse effects on animal and human health [33–36]. On the other hand,

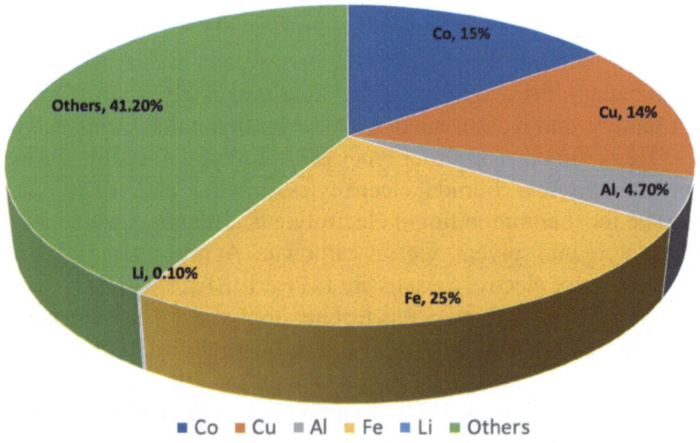

Fig. 3 Metal contents of lithium-ion battery of mobile phones

resources such as cobalt, lithium, iron, copper, manganese, and aluminum in spent lithium-ion batteries (especially cobalt and lithium, which are expensive and scarce resources) can be used again to bring significant economic benefits.

Cobalt has high temperature performance and can be used to manufacture various high-load heat-resistant parts. It can also be used as an additive for acid-resistant alloys and as a binder for hard alloys. The isotope cobalt-60 is an inexpensive source of gamma radiation and has been widely used in the physical, chemical biological research and medical sectors. China's annual demand for cobalt is about 0.06 ~ 0.0800 tons, of which more than 60% need to be imported [37]. In addition to lithium-ion batteries, metallic lithium has been widely used in aerospace materials such as Al–Li alloys and Mg–Li alloys, organic synthesis, tire rubber industry, nuclear fusion reaction power stations, etc., and also has high recycling value [38]. It is estimated that the annual production of waste lithium ion batteries in China is about 67,500 tons, equivalent to 23,000 tons of lithium cobalt acid, lithium nickel cobalt acid/lithium nickel cobalt manganese acid, 6100 tons of copper, and 4100 tons of aluminum. Take mobile phones as an example: the annual sales volume in the country has reached 200 million units, and the value of a mobile phone battery that can be recycled is about 3–4 yuan, with a total value of about 600–800 million yuan [39]. Therefore, resource treatment of spent lithium-ion batteries can not only turn waste into treasure, but also reduce environmental pollution, thereby achieving a win-win situation for environmental protection and economic development [40–43].

Based on the environmental risk and resource characteristics of spent lithium-ion batteries, research on the recycling and utilization of spent lithium-ion batteries has become a global research hotspot. At present, the key and difficult point of recycling lithium-ion batteries is to recover precious metals such as cobalt and lithium from the positive electrodes of the batteries. Commonly used methods are hydrometallurgy, pyrometallurgy and mechanical physics methods.

Hydrometallurgy, usually directly put spent lithium-ion battery anodes in strong acid solution for extraction of cobalt and lithium. After purification of slag during the lithium-ion leaching process, put the corresponding precipitant into cobalt ions to get pure lithium cobalt salt and chemical products (such as cobalt oxalate and lithium carbonate). These chemical products can be used for the preparation of pure state or oxidation state of target products of raw materials such as Co, Li and $LiCoO_2$ [44, 45]. The hydrometallurgical process is gradually mature and efficient, while it is easy to form secondary pollution due to the consumption of large amounts of acid, alkali and precipitant, etc. At the same time, there is a large amount of aluminum dissolved in the leaching solution, resulting in complex purification process and waste of metal aluminum resources.

Pyrometallurgy is to remove the binder on the electrodes by high temperature heating, so as to strip the active materials adhered to the aluminum foils, and then separate and recover the lithium cobalt-rich aggregates and aluminum metals [46, 47]. But the pyrometallurgy process cannot recover organic components, and high energy consumption, easy to produce waste gas, need subsequent treatment of supporting equipment. Mechanical physical method is to use the poor physical characteristics of spent lithium-ion battery components to separate and enrich their component materials by means of crushing and dissociation, air separation, magnetic separation and electrostatic separation, so as to recover the lithium cobalt oxide and aluminum rich collectives [48–50]. However, the recovered lithium cobalt oxide by mechanical and physical methods is still a multi-component mixture, which needs to be refined by hydrometallurgy and other methods to obtain high purity target products. From the perspective of the recycling of lithium cobalt oxide in the lithium-ion battery industry, after the batteries cannot work, the above methods will be used for separation and purification, and then for the preparation of the positive electrodes of lithium-ion batteries. A scientific and effective method can be adopted to directly repair and reuse the failed lithium cobalt oxide, the above process pollution can be greatly reduced, which is of great significance for promoting energy conservation and emission reduction of lithium-ion battery industry and realizing low-carbon economic development.

2 Environmental Risk of Spent Lithium-Ion Battery

Lithium-ion batteries are electronic consumables with an average service life of 1–3 years. As their speed of replacement increases, a large number of spent lithium-ion batteries will be produced. It is estimated that the number of spent lithium-ion batteries in the world will exceed 25 billion in 2020, and the weight will reach more than 500,000 tons. It has been estimated by China's electronic information industry that in 2015, China's spent lithium-ion battery capacity was about 9.4 billion, and it increased to about 11.2 billion in 2016 [51]. With the rapid development of lithium-ion related emerging industries (electric vehicles, smart grids, etc.), the output of China's spent lithium-ion batteries will explode in the future. It has been estimated that, the output of waste spent batteries will increase of 0.07 million tons to 464,000 tons in 2025, with a compound annual growth rate of 59.0% [52]. The huge production of spent lithium-ion batteries will bring potential risks to the environment, and its environmental governance poses a serious challenge [53].

Environmental problems caused by spent lithium-ion batteries have caused widespread concern in countries around the world. Spent lithium-ion batteries contain both original constituents and new substances generated by side reactions during charge and discharge, such as cobalt, lithium, manganese, nickel, organic electrolytes, and additives. Once the spent batteries are discarded in the environment, the physical and chemical reasons are broken and the substances in the battery enter the environment, which cause extremely serious environmental damage and pollution.

$LiCoO_2$ is widely used as a positive electrode material for lithium-ion batteries due to its superior electrochemical performance and stable product performance. However, cobalt metal is a heavy metal while it is free in nature, it will cause great harm to the human body. For example, long-term exposure to cobalt-containing particles may cause inflammation of the upper respiratory tract, contact dermatitis, and gastrointestinal disorders. In addition, once the $LiCoO_2$ burn or decompose by heat, it will undergo redox reaction with water, air, etc. to form toxic and harmful substance, causing environmental pollution and causing an increase in the environmental pH [54]. What's more, metallic nickel as an electrically conductive linker has obvious carcinogenicity, and once dissolved in blood, it can induce lung cancer and sinus cancer, damage the central nervous system, and cause vascular variability [55].

For the anode, graphite carbon materials are generally used as the commercial materials for lithium-ion batteries. Once the graphite carbon materials react with strong oxidants, such as fluorine and liquid chlorine, the combustion can generate CO, CO_2 and other gases. However, the mixture of dust and air may explode when exposed to heat or fire, which will cause dust pollution and increase the pH value of the environment.

$LiPF_6$ and $LiClO_4$, commonly used as electrolytes, are highly corrosive and volatile. Once it is exposed in the air, it will rapidly react with oxygen, water, etc. to form HF, P_2O_5, Li_2O and other toxic substances [56]. The main organic solvents in commercial batteries are propylene carbonate (PC), vinyl carbonate (EC), dimethyl carbonate (DMC), diethyl carbonate (DEC) and methyl ethyl carbonate (EMC). Once these substances come into contact with water, air and strong oxidative reductants, they will react violently, hydrolyze to form substances such as aldehydes, ketones and acids, and burn to form gases such as CO and CO_2, which may even cause explosions. What's more, as for binders, PVDF, VDF and EPD can react with fluorine, fuming sulfuric acid and strong alkali, which can produce HF when heated and cause fluorine pollution. Since the shells are mostly made of steel or aluminum, the shells are difficult to degrade and is exposed to the air for a long time. It will gradually enter the water or soil, and cause long-term and continuous pollution to the environment. Furthermore, most of the separator are made of PP, PE and other organic materials, once combustion, will produce CO, aldehyde, acid and other substances into the environment which resulting in environmental pH value rise and causing serious air pollution. In addition, during the use of lithium-ion batteries, a series of toxic and harmful substances will be produced due to the occurrence of side reactions, such as the production of side reactions between electrolyte and cathode/anode, including HF, CH3OCO2Li, etc., and the decomposition products of organic solvents, such as propylene, ethanol, etc. These substances may directly or indirectly into the environment and trigger environmental pollution.

In summary, arbitrarily discarded spent lithium-ion batteries in environment without reasonable treatment may cause great damage to the environment, including heavy metal pollution, fluorine pollution, dust pollution, air pollution, and water pollution, etc. These pollutions will eventually enter the human body through the enrichment process of "contaminants-soil, water, air-microbes-crops-food-human nervous system", which may lead to a series of serious diseases and complications, even to death.

3 Treatment of Spent Lithium-Ion Battery

3.1 Separation of Electrode Materials for Spent Lithium-Ion Batteries

Based on the structure and material composition of lithium-ion battery, the whole components of spent lithium-ion battery can be recovered by recycling (Fig. 4).

Full component recovery strategy includes: (1) Discharge of spent lithium-ion batteries. (2) Disassembly of spent lithium-ion batteries and classify the components after disassembly. (3) Component separation of electrode materials and metal collector fluid. (4) High value refining of active materials of anode and cathode.

When the battery failure is in the scrapping stage, there is still a certain amount of power remaining. In order to avoid the danger of short circuit and spontaneous combustion, the spent lithium-ion batteries need to be fully discharged [57]. The usual practice is to place it in a salt solution, such as immersing the batteries in a 5% NaCl solution for 24 h, then air drying naturally [58]. After full discharge, the lithium-ion batteries are disassembled manually or mechanically, and the disassembled components are divided into metal and plastic shells, cathode, anode and organic components. Among them, shells and organic membranes with single components and high purity can be directly recycled and treated in a centralized manner.

The active materials applied to both sides of the anode/cathode collector of the lithium-ion battery is a mixture of $LiCoO_2$/graphite, a small amount of a conductive agent, and a binding polyvinylidene fluoride (PVDF) binder. Due to the complex structure and composition, the resource recovery of cathode and anode materials has become the focus and difficulty of research on spent lithium-ion batteries [59–65]. The separation of $LiCoO_2$/graphite and aluminum foil/copper foil are the first step in

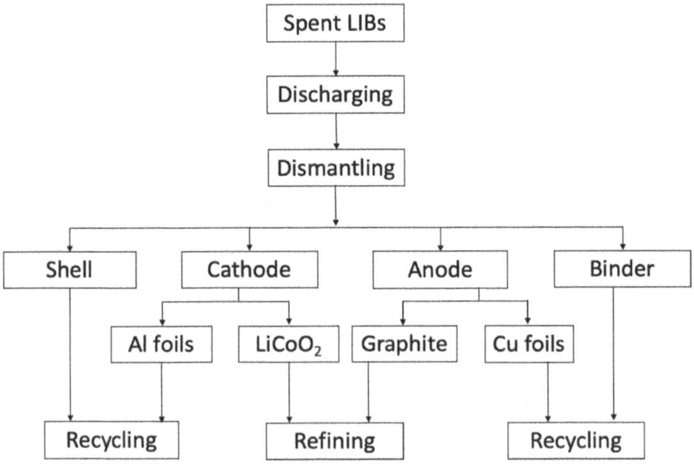

Fig. 4 Schematic diagram of full component recycling process for spent lithium-ion batteries

the recovery of electrode material resources. The binding force between the graphite and the copper foil is relatively weak. After the battery is disassembled, some of the graphite can be separated from the copper foil and recycled directly. Zhu Shuguang et al. [66] separated the anode materials by mechanical pulverization and sieving process after disassembling the spent lithium-ion batteries, and found that the graphite materials were easy to fall off and separate, and the copper recovery rate was more than 93.10 wt%. The adhesion between $LiCoO_2$ and aluminum foils is very strong [67], and it is necessary to separate them by suitable methods. The usual separation methods include mechanical treatment, heat treatment and solvent extraction.

3.1.1 Mechanical Treatment

Although lithium-ion batteries are complex and variable, the large differences in physical properties such as density, ferromagnetism, and electrical conductivity of the constituent materials make it possible to separate and enrich these materials using environmentally friendly mechanical and physical methods. Scientists have conducted fruitful research on this issue. Mechanical sorting method usually firstly crushes and screens the complete electrode materials to obtain the electrode material powder, and then selects the flotation, magnetic separation and the physical properties (such as hydrophilicity, density, magnetic) of the scrap. The specific process flow is shown in Fig. 5.

Zhang et al. [68] used a shear crusher to cut spent lithium-ion batteries into pieces and then pulverized them through an impact crusher with an air-isolated crushing chamber. The pulverized products are mainly divided into three parts: Al enriched component (>2 mm), Cu and Al enriched component (0.25 ~ 2 mm), Co and graphite enriched component (<0.25 mm). After the above treatment, aluminum and copper are separated from the $LiCoO_2$ and graphite material, and can be directly collected and recycled; due to the opposite surface wetting characteristics, $LiCoO_2$ and graphite powder can be separated by flotation method. What's more, since the material properties of anode are quite different, such as copper particles (>0.250 mm) and carbon powder (<0.125 mm) in the pulverized particles, copper and graphite carbon materials can be recovered by methods such as pulverization, vibrating screen and air separation with the recovery rates of 92.4 and 96.6%, respectively [69].

Due to its simple principle and easy to operation, mechanical treatment has been widely used as the mainstream technology for industrial e-waste recycling. At the

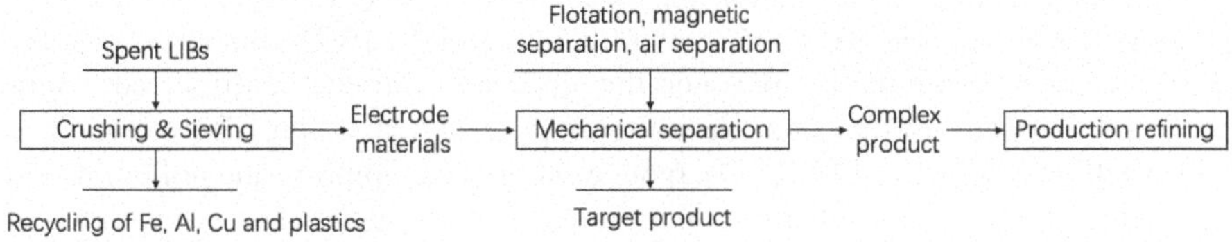

Fig. 5 Typical mechanical sorting process of spent LIBs

same time, since the waste lithium-ion batteries are composed of plurality of materials, the components obtained after the pulverization, separation and enrichment processes are still mixed materials. Further refining process and high value techniques can be combined with mechanical/physical separation methods to improve the efficiency of separation and the quality of the product.

3.1.2 Heat Treatment Method

The heat treatment methods such as calcination and pyrolysis mainly decompose the organic substances by high-temperature incineration to realize separation between constituent materials of the lithium-ion batteries.

Meshram et al. [70] separated the aluminum foil from the cathode sheet at 250 ~ 300 °C for 30 min. Sun et al. [71] found that $LiCoO_2$ powder was almost completely peeled off from the aluminum foil after vacuum pyrolysis at 30 °C for 30 min. When conducting heat treatment on electrode components, the calcination temperature should be controlled carefully. Relative low temperature is hard to effectively remove the PVDF binder and organic additives; the temperature should not be higher than 600 °C to prevent the burning of graphite. The heat treatment method with simple principle and convenient operation is also widely used in the treatment and disposal of e-waste. One thing has to be mentioned that, in the process of calcination, some exhaust gases are inevitably generated due to the combustion of carbon and organic compounds, and the corresponding purification equipment are needed to prevent environmental pollution.

3.1.3 Solvent Extraction Method

Organic solvent methods were based on the principle of "similar phase solution", using strong polarity of the organic solvents to dissolve the electrode binder polyvinylidene fluoride (PVDF), $LiCoO_2$ powder peel off from the aluminum foils, and can eliminate the interference of aluminum elements, so as to simplify the recovery process, improve the recovery effect.

Solvent extraction method can remove PVDF binder to achieve the separation of lithium cobalt oxide/graphite and aluminum foil/copper foil, but it is expensive and highly toxic, and the residue after separation needs to be treated urgently, which restricts the application of solvent extraction method in industry. Moreover, the ultrasonic cleaning utilizes the cavitation and mechanical effect of ultrasonic waves in the liquid to directly and indirectly affect the $LiCoO_2$ and the PVDF binder, so that the PVDF binder between the $LiCoO_2$ and the aluminum foil can be dispersed, emulsified and stripped. So that to achieve the purpose of separation. The method can improve the recovery rate of $LiCoO_2$, reduce energy consumption and pollution, and thus gradually attracts the attention of researchers and is applied to the separation of positive electrode components.

Zhang [72] used NMP (n-methyl-2-pyrrolidone) to dissolve PVDF binder to realize the separation of aluminum foil and cobalt acid lithium/graphite/copper foil at 100 °C temperature. Aluminum foil and copper foil can be directly recycling and NMP can be reuse after vacuum distillation. Li jinhui et al. [44] crushed the waste lithium-ion batteries through a sieve with a diameter of 12 mm, and put the product under the sieve into an ultrasonic cleaner to separate the electrode materials and metal collector fluid. The electrode materials after ultrasonic cleaning were screened through a sieve with a diameter of 2 mm. The product under the sieve was the recyclable electrode material, and the Al foils in the components of 2 ~ 12 mm can be separated directly.

Based on the synergistic effect of PVDF dissolution in organic solvents and ultrasonic cavitation effect mechanism, Yu et al. [73] used ultrasonic to assist organic solvent extraction for the separation and recovery $LiCoO_2$ and Al foils. With the condition of NMP as organic cleaning agents, at 70 °C temperature, ultrasonic power of 240 W, 90 min of reaction time, the recovery rate of $LiCoO_2$ was as high as 99% and the Al foils can be directly recycled at the same time.

3.2 *High-Value Refining Technology*

After separation and enrichment with Al foils/Cu foils, the active electrode materials are mainly $LiCoO_2$ and graphite, and high value-added chemical products can be obtained through refining. Developing scientific and effective high-value refining technology for $LiCoO_2$ and graphite can effectively alleviate the problem of resource shortage and promote the sustainable development of economy.

3.2.1 Mechanochemical Method

The mechanochemical method utilizes a large amount of thermal energy generated by mechanical grinding to push the electrode materials react with the abrasive, thereby converting $LiCoO_2$ into other salts, and then extract them by an appropriate method. For example, Saeki et al. [74] used a planetary ball mill to co-mill $LiCoO_2$ with PVC (polyvinyl chloride) to form chloride salt of cobalt and lithium. Then, based on the solubility of the salts, the cobalt formed by the reaction can be conveniently recycled. The results show that after 30 min of grinding, about 90% of the Co and 100% of the Li reacted with Cl (in PVC) formed inorganic chloride, and almost 90% of the chlorine in the PVC is converted to inorganic chlorine.

The principles of mechanochemical methods are simple, easy to operate, and the use of abrasives can significantly improve the recovery efficiency of Co and Li, and also reducing plastic pollution. However, since the reactants are not sufficiently uniformly contacted, the stoichiometric ratio of the reactants is not easily adjusted, so that the properties of the ground product are unstable. Moreover, mechanochemical methods consume a lot of time and power, which limit their wide application in industry.

3.2.2 Heat Treatment

The research work on the synthesis of high value-added Co and Li products by heat treatment has attracted the attention of researchers. Bahgat et al. [48] heated the electrode materials at 100 ~ 150 °C for 1 h, then added Fe_2O_3 to mix it, and finally heated at 900 ~ 1100 °C for 1 ~ 10 h to synthesize magnetic $Li_{0.5}Fe_{2.5}O_4/CoFe_2O_4$ alloy. Lee et al. [47] first prepared a gel-like precursor by reductive leaching of the electrode materials, and then calcined the precursor at 950 °C for 24 h to successfully obtained pure crystalline $LiCoO_2$. The surface area, charging capacity, and discharge capacity of the material were 30 cm^2 g^{-1}, 165 and 154 mAh g^{-1}, respectively.

Based on the idea of "waste + waste → resources", Xu et al. [75] proposed a new method for in situ anaerobic calcination and wet magnetic separation to recover Co, $LiCoO_2$ and graphite from the mixed electrode materials. This study explored the feasibility of used the negative graphite materials to reduce $LiCoO_2$ of cathode in situ to obtain high value-added products, and provided a theoretical basis for the recovery of the full-component resources of waste lithium-ion batteries. At present, heat treatment is widely used in the industry to synthesize high value-added cobalt/lithium products. However, the waste flue gases generated in the process need further treatment with suitable environmental protection equipment.

3.2.3 Electrochemical Method

Lain [76] introduced the method of battery resource recycling in AEA process at ambient temperature. Co^{3+} can be reduced to Co^{2+} while Li is released from the electrode solid structure. Myoung et al. [77] first leached Co^{2+} from $LiCoO_2$ by nitric acid, and then deposited Co^{2+} as $Co(OH)_2$ on a titanium electrode plate at a constant potential. The deposited $Co(OH)_2$ was at 200 °C to form Co_3O_4. Freitas et al. [78–80] also used electrochemical methods to recover Co from spent lithium-ion batteries, and developed series of discussions on effect of the electrochemical deposition of Co, the change of pH in electrochemical reactions, and the effect of pH on crystal morphology.

The electrochemical method can recover and enrich Co and Li without introducing new impurities and pollution, thereby to obtain high-purity products, which is a relatively environmentally friendly method. However, electrochemical methods consume a lot of time and power, which limits their wide application in industry.

3.2.4 Hydrometallurgy

Due to its low energy consumption, convenient operation, low toxic gas emission and low cost, hydrometallurgical technology has been widely used in the high value industrialization of $LiCoO_2$. Hydrometallurgy first leaches cobalt and lithium metal ions with chemical reagents such as sodium hydroxide, sulfuric acid, nitric acid and hydrogen peroxide, and then separates them by subsequent purification methods such as precipitation, extraction and electrochemistry. The process was shown in Fig. 6.

(1) **Leaching of Co and Li**

① **Acid leaching**

The strong acids used in the leaching experiment are mostly sulfuric acid, nitric acid and hydrochloric acid solutions, and the reaction formula has shown as reactions (1)–(3), and the reaction speed was accelerated by adding H_2O_2 solution.

$$2LiCoO_2(s) + 8HC1 = 2LiCl + 2CoCl_2 + Cl_2(g) + 4H_2O \quad (1)$$

$$4LiCoO_2(s) + 6H_2SO_4 = 2Li_2SO_4 + 4CoSO_4 + O_2(g) + 6H_2O \quad (2)$$

$$4LiCoO_2(s) + 12HNO_3 = 4LiNO_3 + 4Co(NO_3)_2 + O_2(g) + 6H_2O \quad (3)$$

Sakultung et al. [81] compared the leaching effects of cobalt and lithium by sulfuric acid, nitric acid and hydrochloric acid. It was found that HCl had better leaching effect on the two metals under various conditions, and the leaching effect of H_2SO_4 solution was better than HNO_3. The dissolution of $LiCoO_2$ required self-oxidation and reduction reaction, in which Co^{3+} was reduced to Co^{2+}, and O^{2-} was oxidized to O_2. Compared with sulfuric acid and nitric acid, in the absence of reducing agent, the reaction process was difficult while a better leaching effect can be obtained by HCl according to its strong reducibility. In the preferred leaching study, Wang et al. [82] used HCl to dissolve and precipitate the cathode electrode materials. The results showed that the reaction temperature was at 80 °C in 4 M HCl, 1 h of leaching time and 0.02 $g.ml^{-1}$ of solid/liquid ratio, more than 99% of the Co and Li were recovered in the form of $Co(OH)_2$ and Li_2CO_3.

However, when HCl was used for leaching agent, harmful gas like Cl_2 was generated during the reaction, which not only polluted the environment, but also increased

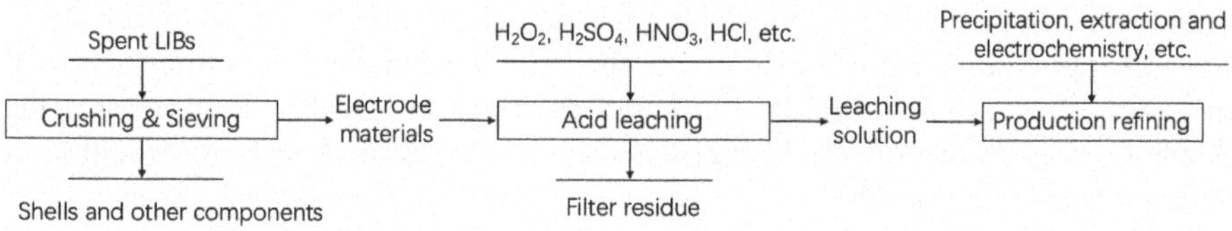

Fig. 6 Hydrometallurgy process of spent LIBs

the requirements for actual production equipment. The most commonly used leaching system is a mixture of H_2SO_4 and H_2O_2.

Ferreira [83], Nan [84], Shin [85] and other research teams have adopted a H_2SO_4 - H_2O_2 leaching system for waste lithium-ion batteries, which had better effects of leaching Li, Co metals. In addition, Jin et al. [86] studied the effect of ultrasonic assisted H_2SO_4 leaching on $LiCoO_2$ electrodes. The results showed that under mechanical agitation, the leaching rate of Co in the electrodes was significantly higher than that without mechanical agitation, because the leaching process controlled by the diffusion process, mechanical agitation accelerated the diffusion of the reaction product in the solution, thereby facilitated the continuation of the reaction, while the ultrasound can generate strong shock waves and jets at a very small area, thereby facilitated the continuation of the reaction. However, this microscopic disturbance has no obvious effect on accelerating diffusion. In addition, when the concentration of H_2SO_4 is less than 1.0 mol L^{-1}, the effect of ultrasonic is significant, while when the concentration of H_2SO_4 is greater than 1.0 mol L^{-1}, the effect of ultrasonic was limited, because it cannot produce reducing agent H_2O_2 in a high concentration solution. It is a good idea to use ultrasonic-assisted methods to improve the conditions during leaching process, which was beneficial to reduce reagent consumption and environmental pollution.

Acidic gas and wastewater are usually produced by using strong acid as leaching agent, which was difficult to handle, pollute the environment, has strong corrosion and high equipment requirements, and also increases production cost to a certain extent. Generally, a large amount of impurity ions was introduced to increase the subsequent separation and purification steps. Therefore, the leaching process currently studied on more selective organic acid as a leaching agent for valuable metals recovery. The organic acid is mild and has almost the same leaching efficiency and reaction conditions as the inorganic acid.

As an environmentally friendly organic acid, citric acid has excellent leaching performance, such as high leaching efficiency, no toxic gas generation, easy natural degradation, relatively low price, etc., which can be used as an excellent acid leaching agent during hydrometallurgy process [87, 88]. Li et al. [87] used citric acid and H_2O_2 as leaching agents to carry out acid leaching reaction of cobalt and lithium. In the batch extraction reactor, in 1.25 mol L^{-1} citric acid solution and 1.0 vol.% H_2O_2, solid-liquid ratio of 20 g L^{-1}, stirring speed of 300 rpm, reaction temperature of 90 °C, and a reaction time of 30 min, more than 90% of Co and about 100% of Li were recovered under the condition.

② Bio-leaching

In the field of metallurgy, bioleaching has gradually gained more and more attention and application, and it is also one of the main development directions of useful metal recycling in spent lithium-ion batteries. The biological method utilizes the metabolic process of microbial fungi with special selectivity to achieve leaching of elements such as Co and Li.

Acidithiobacillus ferrooxidans is a kind of mineralized autotrophic and eosinophilic bacterium that can use energy sources such as elemental S and Fe^{2+}

to produce metabolites such as H_2SO_4 and Fe^{3+} in acid immersion media, thereby contributing to the dissolution of metal elements from spent lithium-ion batteries. Mishra et al. [89] used *Acidithiobacillus ferrooxidans* to leach metal elements in lithium-ion batteries. The results showed that the bio-dissolution of Co is faster than that of Li, and because Fe^{3+} formed coprecipitation with metals in the leachate, higher content of Fe^{2+} would reduce the solubility of the metal; Meanwhile, higher solid/liquid ratio (w/v) would also reduce the solubility of metals, since high metal concentration was harmful to the growth of microbial cells.

Xin et al. [90] explored the combined mechanism of sulfur-oxidizing bacteria and iron-oxidizing bacteria on the leaching of cobalt and lithium in waste lithium-ion batteries, and discussed the methods and process conditions for improving the efficiency of biological dissolution. Lithium precipitation reached the highest when using elemental sulfur as an energy source at a pH of 1.54, and lithium was the lowest when FeS_2 was used as an energy source at pH 1.69. In contrast, the maximum precipitation of Co was occurred at higher pH values and different ORP (redox potentials) when using the $S + FeS_2$ energy system. The above results indicated that no matter what kind of energy system, acid solubility was the main mechanism of lithium precipitation, the acid solubility mechanism of cobalt changed with the energy system. The acid dissolution of the elemental sulfur energy system was the only mechanism of cobalt precipitation, while the combined mechanism of acid dissolution and Fe^{2+} medium reduction in the FeS_2 and $FeS^{2+} + S$ systems played an important role.

In order to meet the future industrial scale, it is important to improve the reaction conditions and recovery efficiency of bioleaching since it takes more time and has relatively low recovery efficiency (Table 1).

(2) **Recovery of Co and Li**

① **Chemical precipitation**

Based on different chemical behavior between cobalt and lithium in different valence states, the precipitation method is generally used to purify and remove impurities in the solution containing cobalt and lithium ions extracted through the acid solution system. Cobalt and lithium salt are precipitated to filter and finally dry the products.

Wang et al. [82] obtained $Co(OH)_2$ by dissolving electrode materials of spent lithium-ion batteries with acid and adding 1 M NaOH solution in the leaching solution to adjust the pH to 11 to precipitate. After $Co(OH)_2$ was obtained, Li_2CO_3 was obtained by adding saturated Na_2CO_3. The results showed that the recovery of Co and Li reached at 96.97 and 96.94%, respectively. After adding $(NH_4)_2C_2O_4$ and Na_2CO_3 into leaching solution, $CoC_2O_4{\cdot}2H_2O$ and Li_2CO_3 were precipitated. 96.3% of Co and 87.5% of Li were dissolved in 2 M H_2SO_4 and 2.0 vol.% H_2O_2 solution, and 94.7% of Co and 71.0% of Li were precipitated in the form of $CoC_2O_4{\cdot}2H_2O$ and Li_2CO_3 [98].

Nan et al. [99] proposed to recover a large amount of Co into cobalt oxalate, and then extracted a small amount of residual cobalt in the solution with Acorga M5640

Table 1 Lithium leaching conditions and efficiency of Co, Li in spent LIBs

Leaching agents	Temperature (°C)	Time (min)	S/L ratio ($g\ L^{-1}$)	Reducing agent	Efficiency	References
4 M HCl	80	60	20	–	Co, Li > 99%	[82]
2.5 M H_2SO_4	60	240	1/10	–	Co: –, Li: 97.2%	[91]
1 M HCl	95	240	1/50	–	Co: 66.2%, Li: 93.4%	[70]
4 M HCl	85	120	1/10	10 vol.% H_2O_2	Co: 95%, Li: 96%	[92]
2 M HNO_3	80	120	–	–	Co: –, Li: 100%	[93]
1 M HNO_3	75	60	20	1.7 vol.% H_2O_2	Co: 85%, Li: 85%	[9]
1.25 M HNO_3	90	30	20	1.0 vol.% H_2O_2	Co: 90%, Li: 100%	[94]
1 M oxalic acid	80	120	50	–	Co, Li > 98%	[95]
1.5 M maleic acid	90	40	20	2.0vol.% H_2O_2	Co: 93%, Li: 94%	[94]
1.25 M ascorbic acid	70	20	25	–	Co: 94.8%, Li: 98.5%	[96]
9 K medium	30	28,800	5	3 g L^{-1}Fe(II), 1% sulpher	Co: 65%, Li: 10%	[89]
Aspergillus niger	30	43,200	1/100	–	Co: 45%, Li: 95%	[97]]

and Cyanex272, respectively. Lithium precipitates in the form of Li_2CO_3, and more than 97% of Co was recovered in the form of cobalt oxalate. Finally, $LiCoO_2$ anode materials with good electrochemical properties were synthesized from recovered cobalt and lithium.

② **Salting out**

This kind of methods is to add some salts into the original solution to be supersaturated, and to precipitate some target solute components, thereby achieving the purpose of recovering valuable metals in the spent lithium-ion batteries. Jin et al. [100] recovered Co from the cathode of spent LIBs by salting out method according to the modern theory of electrolyte solution. When saturated $(NH_4)_2SO_4$ and anhydrous ethanol were added to the HCl solution (which contain $LiCoO_2$), the Co could be salinized under low concentration. When the volume ratio of leaching solution, $(NH_4)_2SO_4$ and anhydrous ethanol was controlled as 2:1:3, the precipitation rate of Co could reach more than 92%. The obtained salted out product was $(NH_4)_2Co(SO_4)_2$ and $(NH_4)Al(SO_4)_2$ which characterized by X-ray diffraction analysis. What's more, Co^{2+} was precipitated from the leachate before Al^{3+}, and different stage of salt separation could separate the two salts in order to obtain different products

③ **Extraction**

Similar to the steps of precipitation method, the extraction method also adopt acid leaching for the first step, while the difference is that Co and Li are separated and recovered by extraction. Therefore, the selection of highly efficient and specific extraction agents is the key in this method. P204, P507, PC-88a and Cyanex 272 are commonly used extraction agents which can ensure relatively high recovery rate and product purity.

After acid leaching of the spent lithium-ion battery using H_2SO_4 and H_2O_2 systems, to avoid the introduction of unnecessary metals, NH_4OH was first added, and about 80% of Co and 95% of Li were leached from the electrode materials. After filtration, the leachate was purified by using Cyanex 272 as an extractant, and about 85% of Co [101] was obtained. After reductive leaching with H_2SO_4 and H_2O_2 system, recovery efficiency of Co was as high as 92% extracted by saponified 0.4 M Cyanex 272 at pH $\approx$ 6 [102].

Precipitation and salting out are safe, economical and have good recycling effects, and have been widely used in industrial production. What's more, solvent extraction can achieve higher recovery and purity. While it is necessary to choose an efficient and economical extractant to prevent the secondary pollution.

The above describes various hydrometallurgical processes. Compared with other methods, these processes are more complicated, a large amount of chemical reagents are consumed, which inevitably cause corrosion of equipment and waste of chemical reagents, and requires corresponding environmental protection. Meanwhile, equipment to eliminate secondary pollution caused by chemical reagents should not be ignored.

3.2.5 Repair and Regeneration of $LiCoO_2$

In summary, hydrometallurgy is the most commonly method for recovering valuable metals in the research of $LiCoO_2$ of spent lithium-ion batteries utilization. From the recycling of $LiCoO_2$ in the lithium-ion battery industry, after they are already to spent, the above methods are used for separation and purification to obtain high value-added products such as $CoC_2O_4{\cdot}2H_2O$ and Li_2CO_3. The preparation of the cathode of the lithium-ion batteries including many steps which involves multiple forms conversions of Co and Li, and the overall process are complicated and the energy consumption is relatively high. If the scientific and effective methods can be used to directly repair and reuse the spent $LiCoO_2$, the above process can be greatly reduced.

(1) **Hydrothermal method**

Compared with common hydrometallurgy, pyrometallurgy and mechanical/physical methods, hydrothermal method can realize the separation and recycling of $LiCoO_2$ in less experiment process. After dismantling, cathode/anode materials of spent lithium-ion batteries can be directly as reactants, without stripping. Under the hydrothermal temperature of 200 °C, Kim et al. [103] placed $LiCoO_2$, Al foil, and separator in a high concentration of LiOH solution, and successfully recovered the regenerated $LiCoO_2$. The reaction was based entirely on the "dissolution-precipitation" mechanism by dissolving the $LiCoO_2$ on the cathode of the spent LIBs. The obtained $LiCoO_2$ has a first discharge capacity of 144.0 mAh g^{-1}, and its discharge capacity can be maintained at 92.2% after 40 cycles, which illustrated a good electrochemical performance.

In addition, Ra et al. [104] used the Etoile-Rebatt (hydrothermal-electrochemical combination process) to regenerate $LiCoO_2$ material by a secondary electrochemical method in a hydrothermal environment. The repaired $LiCoO_2$ has a good layered structure, and the initial discharge capacity reached 134.8 mAh g^{-1}, and can maintain 95.9% after 50 cycles.

The hydrothermal method can achieve the repair of electrochemical performance of $LiCoO_2$, but the reaction requires relatively higher temperature and longer reaction time to meet the energy required for the recovery and transformation of the $LiCoO_2$ crystal structure. Therefore, the hydrothermal repair method has high requirements on the corrosion resistance of the equipment, and thus pose some technical bottlenecks in industrialization.

(2) **Ultrasonic method**

In the practice of repairing active materials of spent lithium-ion batteries, ultrasonic technology is a new method for $LiCoO_2$ regeneration. High-powered ultrasonic waves can create cavitation effects in liquid media and provide a source of energy through cavitation effects, which can improve reaction conditions and help to achieve some difficult chemical reactions [105].

The solid phase $LiCoO_2$ particles and the liquid phase LiOH solution exactly constitute a mass transfer process of Li ions in the ultrasonic environment. The transfer of substances in liquid and solid phases mainly includes crystallization (or dissolution), liquid adsorption (or desorption), leaching and other types.

$LiCoO_2$ has a unique "sandwich structure" whose structure was shown in Fig. 7. $LiCoO_2$ used as cathode materials for lithium-ion batteries has a layered crystal structure of α-$NaFeO_2$ type with R3 m space group, in which O^{2-} is a close-packed cubic core, and Li^+ and Co^{3+} alternately occupy 3a layer of the rock salt structure. At the 3b position, atoms (ions) in the O–Co–O layer are chemically bonded, and the interlayer structure is maintained by van der Waals force. Due to the weak van der Waals force, the presence of lithium ions can just maintain the stability of the layered structure by electrostatic action. The layered structure provides conditions for the migration of lithium.

The excessive release of lithium ions results in the poor lithium in the crystal. During the ultrasonic repair experiment, the system provided the LiOH solution

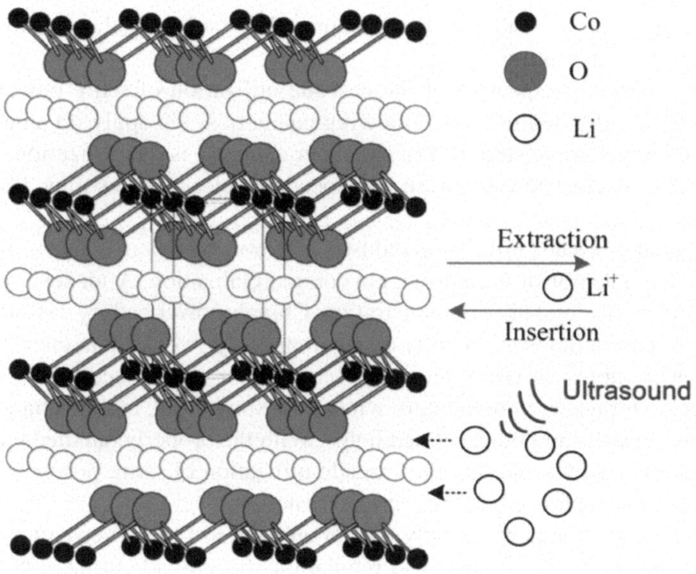

Fig. 7 $LiCoO_2$ crystal structure and ultrasonic mass transfer process

with a lithium-rich environment. It can be seen that there is a difference in the concentration of lithium-ions in the crystal interface of $LiCoO_2$. The concentration of lithium ions in the solid phase is relatively low, while the concentration of lithium ions in the liquid phase is relatively high, which creates environment for mass transfer process. Ultrasound effectively removes the organic matter adhered to the surface of $LiCoO_2$, which also opens the "channel" on the surface of them and provides conditions for the migration of lithium ions. In the experiment, with the help of ultrasonic technology, the strong shear force generated by ultrasonic cavitation in solution and the mechanical disturbance of micro-jet formed pressure changes on the surface of $LiCoO_2$ crystal.

In the process of mass transfer, lithium ions continuously participate in the repair and regeneration of $LiCoO_2$. Meanwhile, in order to obtain $LiCoO_2$ with surplus lithium ions to form new crystal phases, mechanical action of ultrasound can promote the continuous update of the interface, so as to maintain the continuous mass transfer process [105, 106].

4 Cascade Utilization of Spent LIBs

As more and more power lithium-ion batteries are about to be decommissioned, in order to promote the recycling of spent power lithium-ion batteries and promote their continued use in the next life cycle, the establishment of a system for the recycling and utilization of power batteries has become a key factor for the development of power batteries in the future.

From the market perspective of the cascade utilization of retire power battery, it can be roughly divided into the following categories: (1) apply on small energy storage or energy storage station; (2) The other category is the utilization of retired power batteries in electric vehicles in low-speed vehicles, express tricycles, electric bicycle, etc.

At present, the cascade utilization industry of power battery in China is in the transition stage from demonstration project to commercialization. After several years of research, exploration and pilot demonstration, China's power battery cascade utilization and application fields are mainly concentrated in power system energy storage, standby power supply of communication base station, low-speed electric vehicle, family energy storage, complementary wind-solar street light, mobile charging vehicles, electric forklifts and other related fields. With the properly handled technology and the feasible business model, the cascade utilization of retire power battery can be reused and may achieve better economic benefits.

The Chinese government actively encourages the use of power battery cascade utilization, and the policy system and regulations for cascade utilization are gradually improving. From the development history of China's power battery cascade utilization policy:

(1) China first proposed the establishment of power battery cascade utilization management system in the "Energy Conservation and New Energy Automobile Industry Development Plan (2012–2020)";
(2) Relevant policies were introduced between the year of 2012–2016. To this end, the management and legislative direction of power battery cascade utilization were pointed out in China;
(3) In January 2016, the "Technical Policy for Recycling and Utilization of Electric Vehicle Power Battery" was released. The Chinese government has clearly proposed to encourage the development of cascade utilization after battery recycling. Meanwhile, supported the enterprises to continuously develop and innovate the technology of power battery cascade utilization. In February 2016, "Interim Measures for the Management of Standard Conditions and Lists of Cascade Utilization of Waste Energy Batteries for New Energy Vehicles" had been issued by the Ministry of Industry and Information Technology in China;
(4) In February 2018, the "Interim Measures for the Management of Recycling and Utilization of New Energy Vehicles Power Battery" was released, and the use of cascade utilization was clearly stated;
(5) In March 2018, the Ministry of Industry and Information Technology and other seven ministries and commissions carried out pilot projects for the recycling of new energy vehicles, and carried out recycling trials in the Beijing-Tianjin-Hebei, Yangtze River Delta, Pearl River Delta, and Central regions, demanding breakthroughs in the development of the power battery industry which aimed to promote the development of the cascade utilization;

(6) In April 2018, the National Development and Reform Commission's Department of Environmental Protection forwarded the Shenzhen Municipal Government's recycling pilot program had been issued which was the first local government pilot for the recycling of power batteries cascade utilization, taking the use of the cascade as an important content.

Reasonable cascade utilization may bring economic, environmental and social benefits. For the perspective of economic benefits: It extends the life cycle and reduces the cost of power battery, thus bring economic benefits to related enterprises. Meanwhile, rational utilization of spent power lithium-ion batteries can reduce the production of new batteries which reduce energy consumption, carbon dioxide emissions and reduce the risk of environmental pollution. What's more, as for social benefits, cascade utilization for lithium-ion batteries may provide some employment opportunities to the society and may promote the development of social economy to some extends.

References

1. 范亦霞, 我国废旧废弃电子产品回收处理现状及对策研究. 物流科技, (08), 62–64 (2017)
2. N. Armaroli, V. Balzani, Towards an electricity-powered world. Energy Environ. Sci. **4**(9), 3193–3222 (2011)
3. A. Sonoc, J. Jeswiet, A review of lithium supply and demand and a preliminary investigation of a room temperature method to recycle lithium-ion batteries to recover lithium and other materials. Procedia CIRP **15**, 289–293 (2014)
4. 赛迪智库电子信息产业研究所. [2017–08–27]. 锂离子电池产业发展白皮书(2017 版) [EB/OL]
5. J.F. Paulino, N.G. Busnardo, J.C. Afonso, Recovery of valuable elements from spent Li-batteries. J. Hazard. Mater. **150**(3), 843–849 (2008)
6. L. Li, R.J. Chen, F. Sun et al., Preparation of $LiCoO_2$ films from spent lithium-ion batteries by a combined recycling process[J]. Hydrometallurgy **108**(3–4), 220–225 (2011)
7. X.L. Zeng, J.H. Li, Implications for the carrying capacity of lithium reserve in China. Resour. Conserv. Recycl. **80**, 58–63 (2013)
8. C.J. Xu, W.X. Zhang, W.Z. He et al., Generation and management of waste electric vehicle batteries in China. Environ. Sci. Pollut. Res. **24**(26), 20825–20830 (2017)
9. R. Li, Issues and challenges facing rechargeable lithium batteries. Energy **200**, 100 (2001)
10. C.Q. Du, F. Zhang, C.X. Ma, et al., Synthesis and electrochemical properties of $Li_{1.2}Mn_{0.54}Ni_{0.13}Co_{0.13}O_2$ cathode material for lithium-ion battery. Ionics, **22**(2), 209–218 (2016)
11. Y.Y. Xin, X.M. Guo, S. Chen et al., Bioleaching of valuable metals Li Co, Ni and Mn from spent electric vehicle Li-ion batteries for the purpose of recovery. J. Clean. Prod. **116**, 249–258 (2016)
12. C.H. Zhang, L. Yao, Y.P. Qiu, Synthesis and characterization of $LiFePO_4$-carbon nanofiber-carbon nanotube composites prepared by electrospinning and thermal treatment as a cathode material for lithium-ion batteries. J. Appl. Polym. Sci. **133**(9) (2016)
13. D.L. Vu, J.W. Lee, Properties of $LiNi_{0.8}Co_{0.1}Mn_{0.1}O_2$ as a high energy cathode material for lithium-ion batteries. Korean J. Chem. Eng. **33**(2), 514–526 (2016)
14. L.G. Xue, S. Zhang, S.L. Li et al., Synthesis and properties of Li_2MnO_3-based cathode materials for lithium-ion batteries. J. Alloy. Compd. **577**, 560–563 (2013)

15. 姚路.废旧锂离子电池正极材料回收再利用研究: [博士学位论文]. 河南: 河南师范大学 (2016)
16. J. Li, Recycling of spent lithium-ion battery: a critical review. Crit. Rev. Environ. Sci. Technol. **44**(10), 1129–1165 (2014)
17. 何亚群.电子废弃物资源化处理. 北京: 化学工业出版社 (2006)
18. J. Li, C. Daniel, D. Wood, Materials processing for lithium-ion batteries. J. Power Sources **196**(5), 2452–2460 (2011)
19. K.F. Chen, D.F. Xue, Anode performances of mixed $LiMn_2O_4$ and carbon black toward lithium-ion battery. Funct. Mat. Lett. **7**(2) (2014)
20. A.S. Hassan, K. Moyer, B.R. Ramachandran et al., Comparison of storage mechanisms in RuO_2, SnO_2, and SnS_2 for lithium-ion battery anode materials. J. Phys. Chem. C **120**(4), 2036–2046 (2016)
21. X. Zhang, S.C. Han, P.G. Xiao et al., Thermal reduction of graphene oxide mixed with hard carbon and their high performance as lithium ion battery anode. Carbon **100**, 600–607 (2016)
22. F. Wang, Y.Y. Zhang, L.C. Luo et al., Nitrogen-doped carbon nanofiber decorated $LiFePO_4$ composites with superior performance for lithium-ion batteries. Ionics **22**(3), 333–340 (2016)
23. E.M. Wigayati, T. Lestariningsih, A. Subhan et al., Synthesis and characterization of LiBOB as electrolyte for lithium-ion battery[J]. Ionics **22**(1), 43–50 (2016)
24. M.Q. Xu, A. Xiao, W.S. Li et al., Investigation of lithium tetrafluorooxalatophosphate [LiPF4(C2O4)] as a lithium-ion battery electrolyte for elevated temperature performance. J. Electrochem. Soc. **157**(1), A115–A120 (2010)
25. N.L. Hamidah, G. Nugroho, F.M. Wang, Electrochemical analysis of electrolyte additive effect on ionic diffusion for high-performance lithium ion battery. Ionics **22**(1), 33–41 (2016)
26. W.X. Xu, Z.Y. Wang, L.Y. Shi et al., Layer-by-layer deposition of organic-inorganic hybrid multilayer on microporous polyethylene separator to enhance the electrochemical performance of lithium-ion battery. ACS Appl. Mater. Interfaces. **7**(37), 20678–20686 (2015)
27. Y. Lee, M.H. Ryou, M. Seo et al., Effect of polydopamine surface coating on polyethylene separators as a function of their porosity for high-power Li-ion batteries. Electrochim. Acta. **113**, 433–438 (2013)
28. S.J. Gwon, J.H. Choi, J.Y. Sohn et al., Preparation of a new micro-porous poly(methyl methacrylate)-grafted polyethylene separator for high performance Li secondary battery. Nucl. Instrum. Methods Phys. Res. Sect. B-Beam Interact. Mat. Atoms **267**(19), 3309–3313 (2009)
29. K.J. Kim, H.K. Kwon, M.S. Park et al., Ceramic composite separators coated with moisturized ZrO2 nanoparticles for improving the electrochemical performance and thermal stability of lithium ion batteries. Phys. Chem. Chem. Phys. **16**(20), 9337–9343 (2014)
30. S.H. Wang, S.S. Hou, P.L. Kuo et al., Poly(ethylene oxide)-co-poly(propylene oxide)-based gel electrolyte with high ionic conductivity and mechanical integrity for lithium-ion batteries. ACS Appl. Mater. Interfaces. **5**(17), 8477–8485 (2013)
31. 张伟刚.废弃锂离子电池强化选择性破碎研究: [硕士学位论文]. 北京: 中国矿业大学 (2016)
32. 张西华.锂离子电池正极废料中镍钴锰酸锂的短程清洁循环技术: [博士学位论文].北京: 中国科学院 (2015)
33. J.R. Cui, E. Forssberg, Mechanical recycling of waste electric and electronic equipment: a review. J. Hazard. Mater. **99**(3), 243–263 (2003)
34. M. Contestabile, S. Panero, B. Scrosati, A laboratory-scale lithium-ion battery recycling process. J. Power Sources **92**(1–2), 65–69 (2001)
35. G. Chryssos, Safe collection and high value recycling of li-ion batteries. Abstr. Pap. Am. Chem. Soc. **245** (2013)
36. T. Georgi-Maschler, B. Friedrich, R. Weyhe et al., Development of a recycling process for Li-ion batteries. J. Power Sources **207**, 173–182 (2012)
37. M.A. Spreafico, P. Cojocaru, L. Magagnin et al., PVDF latex as a binder for positive electrodes in lithium-Ion batteries. Ind. Eng. Chem. Res. **53**(22), 9094–9100 (2014)

38. 刘全文, 沙景华, 闫晶晶, 等, 中国钴资源供应风险评价与治理研究[J]. 中国矿业, (01), 50–56 (2018)
39. P.K. Choubey, K.-S. Chung, M.-S. Kim et al., Advance review on the exploitation of the prominent energy-storage element Lithium. Part II: From sea water and spent lithium-ion batteries (LIBs). Miner. Eng. **110**, 104–121 (2017)
40. 郭海军, 师增伟, 田歌, 废旧电池回收现状简述[J]. 世界有色金属. (09), 18–20 (2015)
41. 戴长松, 路密, 熊岳平, 等, 废旧锂离子电池处理处置现状及污染防治对策. 环境科学与技术, (S2): 332–335+401 (2013)
42. 陈光. 遵照循环经济的思想促进废旧电池的回收利用; 中国环境科学学会2006年学术年会, 中国江苏苏州, F (2006)
43. 宾智勇, 刘景槐, 吴海国, 等, 废旧锂离子电池的综合利用试验研究. 湖南有色金属, (05), 27–31+80 (2008)
44. J.H. Li, P.X. Shi, Z.F. Wang et al., A combined recovery process of metals in spent lithium-ion batteries. Chemosphere **77**(8), 1132–1136 (2009)
45. C. Shi, Y. Jing, J. Xiao et al., Solvent extraction of lithium from aqueous solution using non-fluorinated functionalized ionic liquids as extraction agents. Sep. Purif. Technol. **172**, 473–479 (2017)
46. K. Tanong, L. Coudert, G. Mercier et al., Recovery of metals from a mixture of various spent batteries by a hydrometallurgical process. J. Environ. Manage. **181**, 95–107 (2016)
47. C.K. Lee, K.I. Rhee, Preparation of $LiCoO_2$ from spent lithium-ion batteries. J. Power Sources **109**(1), 17–21 (2002)
48. M. Bahat, F.E. Farghaly, S.M.A. Basir et al., Synthesis, characterization and magnetic properties of microcrystalline lithium cobalt ferrite from spent lithium-ion batteries. J. Mater. Process. Technol. **183**(1), 117–121 (2007)
49. F. Pagnanelli, E. Moscardini, G. Granata et al., Acid reducing leaching of cathodic powder from spent lithium-ion batteries: Glucose oxidative pathways and particle area evolution. J. Ind. Eng. Chem. **20**(5), 3201–3207 (2014)
50. F. Pagnanelli, E. Moscardini, P. Altimari et al., Cobalt products from real waste fractions of end of life lithium-ion batteries. Waste Manage. **51**, 214–221 (2016)
51. 李飞.废锂电池资源化技术及污染控制研究: [硕士学位论文].四川: 西南交通大学 (2017)
52. 侯兵.电动汽车动力电池回收模式研究:[硕士学位论文]. 重庆: 重庆理工大学 (2015)
53. E. Gratz, Q. Sa, D. Apelian et al., A closed loop process for recycling spent lithium ion batteries. J. Power Sources **262**(4), 255–262 (2014)
54. 李洪枚, 姜亢. 废旧锂离子电池对环境污染的分析与对策. 上海环境科学. (05), 201–203 (2004)
55. 张明, 邵婷, 姬玉玲等. 废旧电池的危害及防治对策. 安全与环境工程. (02), 103–106 (2008)
56. 刘元龙, 戴长松, 贾铮等. 废旧锂离子电池电解液的处理技术. 电池. (02): 124–126 (2014)
57. S. Zhu, Y. Qin, J. Zhang, Renovation of lithium cobalt oxide from spent lithium ion batteries by an aqueous pulsed discharge plasma. Int. J. Electrochem. Sci. **11**(7), 6403–6411 (2016)
58. J. Li, G. Wang, Z. Xu, Generation and detection of metal ions and volatile organic compounds (VOCs) emissions from the pretreatment processes for recycling spent lithium-ion batteries. Waste Manage. **52**, 221–227 (2016)
59. N.J. Boxall, N. Adamek, K.Y. Cheng et al., Multistage leaching of metals from spent lithium ion battery waste using electrochemically generated acidic lixiviant. Waste Manage. **74**, 435–445 (2018)
60. Z. Cao, X. Zheng, H. Cao et al., Efficient reuse of anode scrap from lithium-ion batteries as cathode for pollutant degradation in electro-Fenton process: Role of different recovery processes. Chem. Eng. J. **337**, 256–264 (2018)
61. H. Gomaa, M.A. Shenashen, H. Yamaguchi et al., Extraction and recovery of Co2 + ions from spent lithium-ion batteries using hierarchical mesosponge gamma-Al2O3 monolith extractors. Green Chem. **20**(8), 1841–1857 (2018)
62. X. Chen, C. Guo, H. Ma et al., Organic reductants based leaching: a sustainable process for the recovery of valuable metals from spent lithium ion batteries. Waste Manage. **75**, 459–468 (2018). (New York, NY)

63. Y. Chen, N. Liu, F. Hu et al., Thermal treatment and ammoniacal leaching for the recovery of valuable metals from spent lithium-ion batteries. Waste Manage. **75**, 469–476 (2018). (New York, NY)
64. W. Gao, C. Liu, H. Cao et al., Comprehensive evaluation on effective leaching of critical metals from spent lithium-ion batteries. Waste Manage. **75**, 477–485 (2018). (New York, NY)
65. W. Gao, J. Song, H. Cao et al., Selective recovery of valuable metals from spent lithium-ion batteries - Process development and kinetics evaluation. J. Clean. Prod. **178**, 833–845 (2018)
66. S. Zhu, W. He, G. Li, X. Zhou, J. Huang, X. Zhang, in *Recovering copper from spent lithium ion battery by a mechanical separation process*. 2011 International Conference on Materials for Renewable Energy & Environment,Shanghai: 1008–1012 (2011)
67. T. Zhang, Y.Q. He, L.H. Ge et al., Characteristics of wet and dry crushing methods in the recycling process of spent lithium-ion batteries. J. Power Sources **240**, 766–771 (2013)
68. T. Zhang, Y.Q. He, F.F. Wang et al., Chemical and process mineralogical characterizations of spent lithium-ion batteries: an approach by multi-analytical techniques. Waste Manage. **34**(6), 1051–1058 (2014)
69. 周旭, 朱曙光, 次西拉姆, 等, 废锂离子电池负极材料的机械分离与回收. 中国有色金属学报. (12), 3082–3086 (2011)
70. P. Meshram, B.D. Pandey, T.R. Mankhand, Recovery of valuable metals from cathodic active material of spent lithium ion batteries: Leaching and kinetic aspects. Waste Manage. **45**, 306–313 (2015)
71. L. Sun, K. Qiu, Vacuum pyrolysis and hydrometallurgical process for the recovery of valuable metals from spent lithium-ion batteries. J. Hazard. Mater. **194**, 378–384 (2011)
72. X.H. Zhang, Y.B. Xie, X. Lin et al., An overview on the processes and technologies for recycling cathodic active materials from spent lithium-ion batteries. J. Mater. Cycles Waste Manage. **15**(4), 420–430 (2013)
73. L.P. He, S.Y. Sun, X.F. Song et al., Recovery of cathode materials and Al from spent lithium-ion batteries by ultrasonic cleaning. Waste Manage. **46**, 523–528 (2015)
74. S. Saeki, J. Lee, Q.W. Zhang et al., Co-grinding LiCoO2 with PVC and water leaching of metal chlorides formed in ground product. Int. J. Miner. Process. **74**, S373–S378 (2004)
75. J. Li, G. Wang, Z. Xu, Environmentally-friendly oxygen-free roasting/wet magnetic separation technology for in situ recycling cobalt, lithium carbonate and graphite from spent LiCoO2/graphite lithium batteries. J. Hazard. Mater. **302**, 97–104 (2016)
76. M.J. Lain, Recycling of lithium ion cells and batteries. J. Power Sources **97–8**, 736–738 (2001)
77. J. Myoung, Y. Jung, J. Lee et al., Cobalt oxide preparation from waste LiCoO2 by electrochemical-hydrothermal method. J. Power Sources **112**(2), 639–642 (2002)
78. M.B.J.G. Freitas, E.M. Garcia, Electrochemical recycling of cobalt from cathodes of spent lithium-ion batteries. J. Power Sources **171**(2), 953–959 (2007)
79. V.G. Celante, M.K. Pietre, M.B.J.G. Freitas, Electrochemical and structural characterization of copper recycled from anodes of spent Li-ion batteries. J. Appl. Electrochem. **40**(1), 23–28 (2009)
80. M.B.J.G. Freitas, V.G. Celante, M.K. Pietre, Electrochemical recovery of cobalt and copper from spent Li-ion batteries as multilayer deposits. J. Power Sources **195**(10), 3309–3315 (2010)
81. S. Sakultung, K. Pruksathorn, M. Hunsom, Simultaneous recovery of valuable metals from spent mobile phone battery by an acid leaching process. Korean J. Chem. Eng. **24**(2), 272–277 (2007)
82. R.C. Wang, Y.C. Lin, S.H. Wu, A novel recovery process of metal values from the cathode active materials of the lithium-ion secondary batteries. Hydrometallurgy **99**(3–4), 194–201 (2009)
83. D.A. Ferreira, L.M.Z. Prados, D. Majuste et al., Hydrometallurgical separation of aluminium, cobalt, copper and lithium from spent Li-ion batteries. J. Power Sources **187**(1), 238–246 (2009)

84. J.M. Nan, D.M. Han, M.J. Yang et al., Recovery of metal values from a mixture of spent lithium-ion batteries and nickel-metal hydride batteries. Hydrometallurgy **84**(1–2), 75–80 (2006)
85. S.M. Shin, N.H. Kim, J.S. Sohn et al., Development of a metal recovery process from Li-ion battery wastes. Hydrometallurgy **79**(3–4), 172–181 (2005)
86. 金玉健，梅光军，李树元，废锂离子电池LiCoO2电极中钴的超声辅助浸出．湿法冶金, (02), 97–99 (2006)
87. L. Li, J. Ge, F. Wu et al., Recovery of cobalt and lithium from spent lithium ion batteries using organic citric acid as leachant. J. Hazard. Mater. **176**(1–3), 288–293 (2010)
88. X. Chen, T. Zhou, Hydrometallurgical process for the recovery of metal values from spent lithium-ion batteries in citric acid media. Waste Manage. Res. **32**(11), 1083–1093 (2014)
89. D. Mishra, D.J. Kim, D.E. Ralph et al., Bioleaching of metals from spent lithium ion secondary batteries using Acidithiobacillus ferrooxidans. Waste Manage. **28**(2), 333–338 (2008)
90. B.P. Xin, D. Zhang, X. Zhang et al., Bioleaching mechanism of Co and Li from spent lithium-ion battery by the mixed culture of acidophilic sulfur-oxidizing and iron-oxidizing bacteria. Biores. Technol. **100**(24), 6163–6169 (2009)
91. R. Zheng, L. Zhao, W. Wang et al., Optimized Li and Fe recovery from spent lithium-ion batteries via a solution-precipitation method. Rsc Adv. **6**(49), 43613–43625 (2016)
92. L. Chen, X.C. Tang, Y. Zhang et al., Process for the recovery of cobalt oxalate from spent lithium-ion batteries. Hydrometallurgy **108**(1–2), 80–86 (2011)
93. S. Castillo, F. Ansart, C. Laberty-Robert et al., Advances in the recovering of spent lithium battery compounds. J. Power Sources **112**(1), 247–254 (2002)
94. X.L. Sun, X.H. Wang, N. Feng et al., A new carbonaceous material derived from biomass source peels as an improved anode for lithium ion batteries. J. Anal. Appl. Pyrol. **100**, 181–185 (2013)
95. L. Sun, K.Q. Qiu, Organic oxalate as leachant and precipitant for the recovery of valuable metals from spent lithium-ion batteries. Waste Manage. **32**(8), 1575–1582 (2012)
96. L. Li, J. Lu, Y. Ren et al., Ascorbic-acid-assisted recovery of cobalt and lithium from spent Li-ion batteries. J. Power Sources **218**, 21–27 (2012)
97. N.B. Horeh, S.M. Mousavi, S.A. Shojaosadati, Bioleaching of valuable metals from spent lithium-ion mobile phone batteries using *Aspergillus niger*. J. Power Sources **320**, 257–266 (2016)
98. S.G. Zhu, W.Z. He, G.M. Li et al., Recovery of Co and Li from spent lithium-ion batteries by combination method of acid leaching and chemical precipitation. Trans. Nonferrous Met. Soc. China **22**(9), 2274–2281 (2012)
99. J.M. Nan, D.M. Han, X.X. Zuo, Recovery of metal values from spent lithium-ion batteries with chemical deposition and solvent extraction. J. Power Sources **152**(1), 278–284 (2005)
100. 金玉健, 梅光军, 李树元, 盐析法从锂离子电池正极浸出液中回收钴盐的研究. 环境科学学报, (07): 1122–1125 (2006)
101. G. Dorella, M.B. Mansur, A study of the separation of cobalt from spent Li-ion battery residues. J. Power Sources **170**(1), 210–215 (2007)
102. J. Kang, G. Senanayake, J. Sohn et al., Recovery of cobalt sulfate from spent lithium ion batteries by reductive leaching and solvent extraction with Cyanex 272. Hydrometallurgy **100**(3–4), 168–171 (2010)
103. D.S. Kim, J.S. Sohn, C.K. Lee et al., Simultaneous separation and renovation of lithium cobalt oxide from the cathode of spent lithium ion rechargeable batteries. J. Power Sources **132**(1–2), 145–149 (2004)
104. D.I. Ra, K.S. Han, Used lithium ion rechargeable battery recycling using Etoile-Rebatt technology. J. Power Sources **163**(1), 284–288 (2006)
105. S.G. Zhu, W.Z. He, G.M. Li et al., Renovation of lithium cobalt oxide from the cathode of spent lithium ion batteries by ultrasonic irradiation. Fresenius Environ. Bull. **21**(7), 1752–1757 (2012)
106. 朱曙光. 废锂离子电池中钴酸锂材料超声修复研究 (2012)
107. https://baijiahao.baidu.com/s?id=1593256955044073936&wfr=spider&for=pc

108. http://www.elecfans.com/d/840179.html
109. https://www.xianjichina.com/special/detail_329718.html

Hydrometallurgically Recycling Spent Lithium-Ion Batteries

Jiexi Wang and Zhihao Guo

1 Introduction

Hydrometallurgical methods for recycling spent lithium-ion batteries (LIB) are the most major approaches for recycling spent LIBs since more than half of the recycling processes reported are hydrometallurgical processes [1]. Compared with pyrometallurgical process, hydrometallurgical process embraces a variety of advantages, such as high recycling efficiency, high metal selectivity, low energy consumption, little hazardous gas emission, and low capital cost, which shows huge potential in industrial realization [2]. Although some challenges still exist in these strategies, including the complex operation steps and waste water emission, it is well known that hydrometallurgical methods have a greater potential than the pyrometallurgical methods to realize sustainable development of this industry. Typically, hydrometallurgical process consists of leaching (e.g. acid leaching, alkaline leaching, etc.), purification (e.g. solvent extraction, chemical precipitation, electrochemical deposition, etc.) and recovery procedures as illustrated in Fig. 1.

2 Leaching

Leaching is a key and indispensable step for recovering valuable resources from spent LIBs in the whole hydrometallurgical processes. The aim of leaching is to make the valuable metals and other substances separate and exist in different forms. Leaching is generally used to dissolve metals in the spent materials from the solid state into solution for further processing, which is very similar to other metallurgical processes. Leaching can be also regarded as a pre-treatment procedure before the

J. Wang (✉) · Z. Guo
School of Metallurgy and Environment, Central South University, Changsha, Hunan 410083, China
e-mail: wangjiexikeen@csu.edu.cn

L. An (ed.), *Recycling of Spent Lithium-Ion Batteries*,
https://doi.org/10.1007/978-3-030-31834-5_2

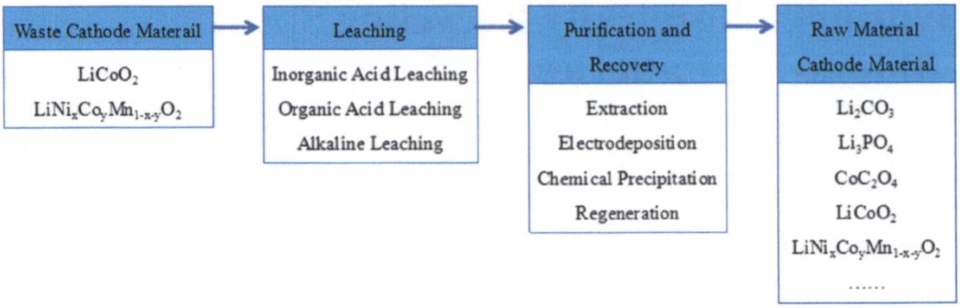

Fig. 1 Hydrometallurgical processes for recycling spent LIBs

subsequent purification and separation process. Thus, the overall recycling rate of metal is greatly affected by the leaching efficiency.

The leaching process of cathode active materials is usually carried out via using inorganic acid, organic acid, alkaline as leaching media. The acid leaching has attracted more attention than the alkaline leaching due to its higher leaching efficiency. Parameters such as leachant concentration, temperature, reaction time and solid-to-liquid ratio (S/L) all have a great influence on the leaching efficiency, and need to be optimized. And measures like ultrasonic wave vibration, mechanical agitation are also employed to help enhance leaching. In this section, we will review some progress about inorganic acid leaching, organic acid leaching, and alkaline leaching of cathode materials from spent LIBs.

2.1 Leaching Mechanism

Understanding the mechanism of leaching is of great importance to help optimize operation conditions. The overall mechanism is that the crystal lattice of cathode materials is broken down in the right leaching media. During the leaching process, hydrogen ion can extract lithium from the cathode materials and its reaction efficiency is independent. Based on the relationship between the acid concentration and leaching efficiency, the profile of leaching performance for the various acids is illustrated in Fig. 2 [1]. It is demonstrated that leaching efficiency is changed from poor to strong with poor acid to strong acid. Leaching efficiency is highly related to the acid type and concentration. It can be also concluded that the concentration of hydrogen ion or pH value is directly related to the leaching efficiency.

However, research about detailed leaching mechanism has rarely been reported. For example, Billy et al. explored the dissolution mechanisms of $LiNi_{1/3}Mn_{1/3}Co_{1/3}O_2$ in acid solution [3]. They proposed a two-steps mechanism: the first step is "self-regulating" by the lithium de-intercalation, the charge compensation of transition metals, and partially by the oxygen reaction, in which this oxygen reaction will induce the release of O_2 gas and the defects formation in the material. The process is highly related to the proton concentration and certainly initiated by an

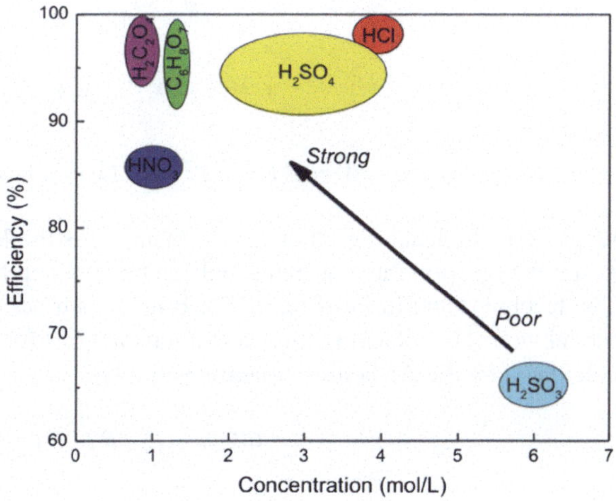

Fig. 2 Comparison of the different acid-leaching technologies in terms of concentration and efficiency [1]

excess of positive charge at the solid/liquid interface of the particles. The delithiation process leads to a positive shift of particles, which reduces the leaching driving force and limits the dissolution. Secondly, the concentration of the surface vacancies that associate to the high potential of particles will lead to a surface reorganization from the layered structure to the metastable birnessite phase or a transformation to the γ-type manganese oxide. This step results from the disproportion reaction of manganese and/or the redox reaction between the Mn^{4+}/Mn^{2+}. It induces an enrichment in manganese at the particle surface, forming a well-defined core-shell structure. In the second step, the dissolution is controlled by the presence of a divalent manganese in solution. Although whether this mechanism is universally true still needs confirmation, this may inspire the new strategies for recycling spent lithium-ion batteries in acid solutions.

In general, thermodynamics calculation of the Gibbs free energy change ($\Delta_r G$) can be used to explore reaction trends and possible leaching products. For instance, the Pourbaix diagrams, also known as Eh–pH diagrams (Eh: the voltage potential with respect to the standard hydrogen electrode), of M–H_2O (M = Li, Ni, Co, and Mn) system can used to predict the stable regions of different phases in aqueous solutions [4]. Different characterization methods, like XRD, SEM, FTIR spectroscopy, Raman spectroscopy, and ultraviolet-visible spectroscopy have been used to confirm the leaching mechanism [5, 6].

In the research of kinetics of leaching process, four leaching models, such as the shrinking core [7], empirical [8], Avrami equation [9], and revised cubic rate law models [10], have been proposed to analyze the leaching kinetics of different metals. The leaching process from the spent LIB cathode materials can be regarded as a solid–liquid heterogeneous process including a series of mass transfer, ion diffusion, and surface chemical reactions. Therefore, based on the shrinking core model, the leaching rate can be assumed to controlled by the liquid film mass transfer (Eq. (1)), the surface chemical reaction (Eq. (2)), or residue layer diffusion (Eq. (3)):

$$X = k_1 \cdot t \quad (1)$$

$$1 - (1 - X)^{1/3} = k_2 \cdot t \quad (2)$$

$$1 - 3(1 - X)^{2/3} + 2(1 - X) = k_3 \cdot t \quad (3)$$

where X is the leaching efficiency of metals, t is the leaching time, and k_1, k_2, and k_3 are the reaction rate constants and can be obtained from the slopes of fitted plots. The leaching data in most reports also fit the surface chemical reaction controlled model well. The relatively high activation energies for the leaching of metals can be calculated by the Arrhenius equation (Eq. (4)).

$$\text{In}k = -E_a/RT + C \quad (4)$$

where k is the reaction rate constants, A is the pre-exponential factor, E_a is the apparent activation energy, R is molar gas constant. and T is the absolute temperature. In other specific cases, the empirical (Eq. (5)), Avrami equation (Eq. (6)), or revised cubic rate law (Eq. (7)) models match leaching data better than the shrinking core model, implying a surface layer diffusion of the lixiviant or a surface chemical reaction-controlled process.

$$(-\ln(1 - X))^2 = k \cdot t \quad (5)$$

$$-\ln(1 - X) = k \cdot t^n \quad (6)$$

$$(1 - X)^{1/3} = 1 - kt/\rho r_0 \quad (7)$$

where n is the suitable parameter, r_0 is the radius of the particle at $t = 0$, and ρ is the density of the particle (Fig. 3).

2.2 *Inorganic Acid Leaching*

Inorganic acid leaching is a convenient method that has been reported massively. In the early stage of research for acid leaching, a few kinds of strong inorganic acids, such as sulfuric acid (H_2SO_4), nitrate acid (HNO_3) and hydrochloric acid (HCl) are the most commonly used leaching agents for the leaching of cathode materials from spent LIBs. The inorganic acid leaching has been proven to be feasible and effective, but the disadvantages like the emission of secondary pollutants and complexity of separation and purification steps have been also appeared.

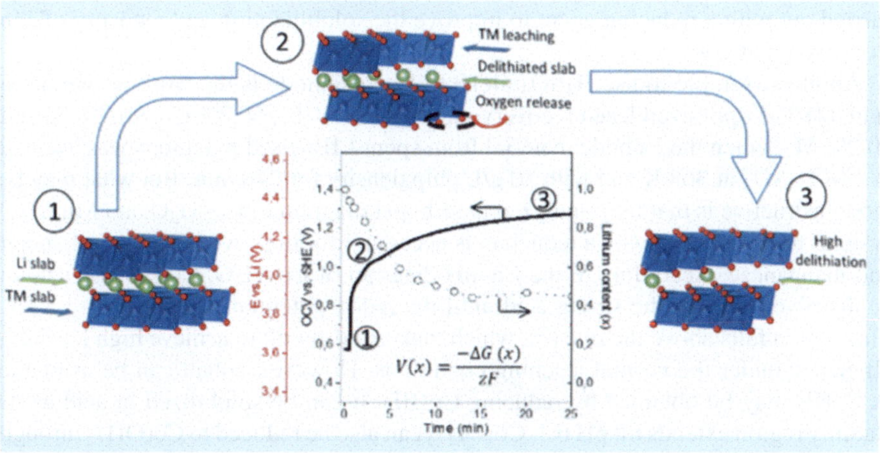

Fig. 3 Schematic illustration of leaching mechanism of NCM in acid solutions [3]

2.3 H_2SO_4 Leaching

$\mathbf{H_2SO_4}$ is the most commonly utilized inorganic acid that has been reported to dissolve cathode materials. The chemical reaction equation is as follows:

$$4LiCoO_2 + 6H_2SO_4 \rightarrow 2Li_2SO_4 + 4CoSO_4 + 6H_2O + O_2 \uparrow \tag{8}$$

Nan et al. [11] utilized H_2SO_4 alone to leach the pretreated LCO cathode material and almost 98% Co and Li were leached out while the dissolution ratio of copper (foil) was only 9% in 3 M H_2SO_4 at 70 °C and with S/L = 1:5 for 6 h. Although the leaching efficiency of cobalt was favorable at a higher acid concentration and a reaction temperature, the reduction ability of sulfuric acid was increased and thus more copper foil would be dissolved. Thus, in order to reach a both satisfying leaching efficiency and separation effect, operational conditions like temperature and acid concentration need to be carefully optimized.

To identify the influence of several parameters of H_2SO_4 Leaching (acid concentration, solid/liquid ratio, leaching time and number of reaction steps) on the removal of metals from spent batteries, a Box-Behnken design was used as reported by Tanong et al. [12] Based on the results, it is seemed that the solid/liquid ratio and acid concentration are the main parameters influencing the solubilization of zinc, manganese, nickel, cadmium and cobalt from spent batteries. The leaching conditions were optimized to maximize the solubilization of metals and to minimize the costs related to the consumption of chemical products. The best conditions were defined as three 30 min-leaching steps carried out in 1 M sulfuric acid, with a solid/liquid ratio fixed at 18% (w/w). Under these optimal conditions, more than 86% of Cd, 70% of Co, 30% of Mn, 70% of Ni and 100% of Zn were solubilized from the mixture of spent LIB cathode batteries. Authors also confirmed that further experiments should be carried out with a reducing agent to improve the solubilization and recovery of Mn from spent batteries.

Another example to use H_2SO_4 alone to leach cathode is the work of Meshram et al. [4] The optimized leach recovery rate was 93.4% Li, 66.2% Co, 96.3% Ni and 50.2% Mn, when the cathode material from spent LIBs used in laptops was leached in 1 M H_2SO_4 at 368 K and with 50 g/L pulp density for 240 min. But what may be more instructive is that they analyzed the Eh–pH diagram of Co–H_2O and Mn–H_2O systems and explained why a reductant is necessary for improved recovery of cobalt and manganese. According to the Eh–pH diagram of Co–H_2O, Co^{3+} phase cannot be dissolved even in the strong acid until the redox potential reaches to ~+1.84 V. This region falls above the line (b), which makes it difficult to achieve high leaching efficiency under the normal leaching conditions. However, cobalt can be available as Co(II) may be obtained by reducing Co(III). It can be solubilized in acid as its stability region extends till pH 6.3. $Co(OH)_3$ can also be reduced to $Co(OH)_2$ through Co_3O_4 in alkaline solution at the higher pH and can be reductively leached (Co^{2+}) in an acid. In the Eh–pH diagram of Mn–H_2O, the domain of stable Mn phases is in the stability region of water, whose lower limit is indicated by line (a). Mn(II) can thus be dissolved in entire acidic region. In order to dissolve Mn(IV) (MnO_2) phase which is also a part of $Li_2CoMn_3O_8$ [a spinel structure containing Mn(III) and Mn(IV)] in the spent batteries, a very strong reducing condition and strong acid solutions are required to form soluble Mn(II) phase. However, at pH > 3.2, Mn(IV) can be solubilized as Mn(II) under relatively lower redox conditions through the formation of an intermediate phase Mn_2O_3(s) (Figs. 4 and 5).

In order to effectively leach cathode materials, adding reductant H_2O_2 to H_2SO_4 solution has been massively reported [13]. In addition to acids, reducing agents also

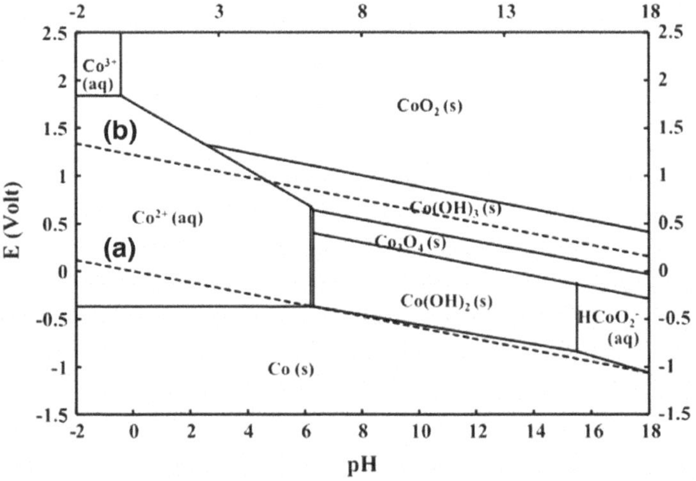

Fig. 4 Eh–pH diagram for Co–H_2O system [soluble species concentration (except H^+) = 0.5 M at 298 K] [4]

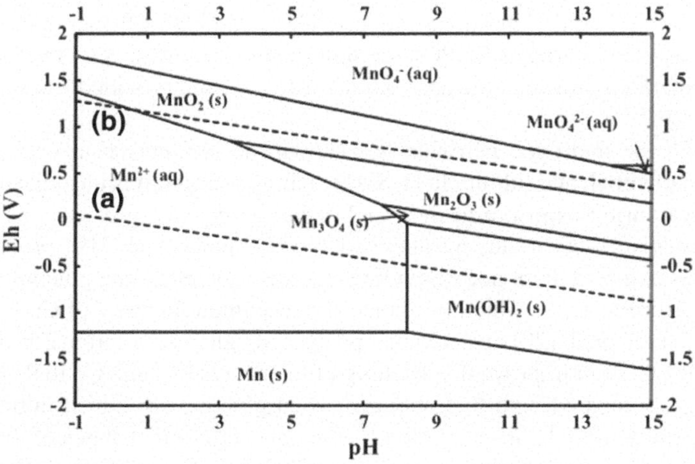

Fig. 5 Eh–pH diagram for Mn–H_2O system at 298 K [soluble species concentration (except H^+) = 0.5 M at 298 K] [4]

play a key role in improving leaching efficiencies. The metals in lower valence states dissolve more readily. For instance, Co^{2+} is much more water soluble than Co^{3+}. The reaction equation is as follows:

$$2LiCoO_2 + 3H_2SO_4 + H_2O_2 \rightarrow 2CoSO_4 + Li_2SO_4 + 4H_2O + O_2 \uparrow \quad (9)$$

Zhu et al. [13] reported that 96.3% of Co and 87.5% of Li can be dissolved in the solution of 2 mol/L H_2SO_4 and 2.0% H_2O_2 (volume fraction) at 60 °C with 33 g/L S:L ratio for 2 h. Jha et al. [14] utilized 2 M sulfuric acid with the addition of 5% H_2O_2 (v/v) to leach $LiCoO_2$ at a pulp density of 100 g/L and 75 °C for 1 h with a recovery of 99.1% Li and 70.0% Co. While in Swian et al.'s work [15], 93% of cobalt and 94% of lithium can be leached under optimized experimental conditions: 100 g/L pulp density, 2 M concentration of leachant, 5% concentration of reductant H_2O_2, 75 °C and 30 min leaching time. It seems that under similar operational conditions, the leaching results are almost same among different reports. Kang et al. [16] used 2 M H_2SO_4 and 6 vol.% of H_2O_2 to leach the spent cathode material at 60 °C for 1 h, and the obtained high leaching efficiency is 98% for Co and 97% for Li. Chen et al. [17] reported that the leaching efficiency of cobalt was 95%, and that of lithium was 96% under optimum conditions that are liquid/solid ration of 10:1, leaching time of 120 min and a temperature of 85 °C in 4 M H_2SO_4 + 10% v/v H_2O_2 mixed solution.

Apart from the cobalt-containing cathode materials such as LCO and NCM, the acid leaching strategy to recovery another commercialized cathode-$LiFePO_4$ using the H_2SO_4 and H_2O_2 mixed solution has been also proposed. Recently, Li et al. [18] developed a selective leaching process by using stoichiometric H_2SO_4 and hydrogen peroxide (H_2O_2) as an oxidant to leach Li from spent LFP; Fe and P remained in the residue as $FePO_4$. The leaching efficiencies of Li, Fe, and P in this process were 96.85, 0.027, and 1.95%, respectively. Then, 95.66% of the Li in the leaching solution was precipitated as Li_3PO_4 through the reaction with Na_3PO_4. This novel method remarkably lowered the amount of waste acid and provided a new approach to recycle LFP.

Although the hydrogen peroxide has shown the advantages in promoting the dissolution of cobalt and lithium in H_2SO_4 leaching process, it is unstable and easily decomposes during exposure to light and/or heat. Recently, glucose was explored as an alternative green reducing agent to H_2O_2. Meng et al. [19] proposed that glucose was oxidized to monocarboxylic acids analogy, including gluconic, tartaric, oxalic, and formic acids, which promoted the leaching efficiency of Co from spent LCO. Pagnanelli et al. [20] revealed that postponing glucose addition during H_2SO_4 leaching of LCO can improve the leaching efficiencies of Li and Co to 92 and 88%, respectively, compared with that obtained when glucose was added initially (60% for both metal ions). The investigation on the concentration changes of the glucose intermediates indicated that this was caused by the shift of the oxidative pathways at different leaching times. When glucose was added at the beginning of leaching process, the intermediate arabic acid was formed and accumulated, slowing down the oxidation process. In contrast, if glucose was added after leaching for 2 h, the reductive intermediate glyoxylic acid was formed, which increased the reducibility of glucose. The two different reaction pathways via glucose are illustrated in Fig. 6.

Joulie et al. [21] recently examined the possibility of Cu and Al current collectors as reducing agents in the leaching of NCM. Cu and Al were chosen because they are usually found in spent cathode materials after the pre-treatment. The results exhibited that high leaching efficiencies of >90% were achieved with 1 M H_2SO_4 at 30 °C in the presence of Cu with an NCM/Cu ratio of 1:1.2 (w/w) or in the presence of Al with an NCM/Al ratio of 1:0.7 (w/w). These novel reducing agents showed environmentally friendly features, such as the low energy demand and toxic gas emissions.

A very innovative idea to reduce spent cathode material was reported by Yang et al. [22]. Instead of using H_2O_2 as the reductant, they made the best of all the components in spent LIBs. The graphite from the spent LIBs was utilized to reduce LCO cathode material before H_2SO_4 acid leaching. To confirm the feasibility of reducing LCO by graphite, they calculated thermodynamics data as shown in Table 1 based on the equation:

$$\Delta \mathrm{G(T)} = \sum_{P} V_P \Delta_f H^{\circ}(P, 298\ \mathrm{K}) - T \sum_{P} V_P \Delta_f S^{\circ}(P, 298\ \mathrm{K}) \qquad (10)$$

where, P represents reactant or product; V_p is the stoichiometric number of P with negative sign for product or positive sign for reactant; $\Delta_f H^{\circ}(P, 298\ \mathrm{K})$ and $\Delta_f S^{\circ}(P, 298\ \mathrm{K})$ are the corresponding standard formation enthalpy and entropy of P at 298 K, respectively. The involved standard formation enthalpy and entropy are from data in Metallurgical Thermochemistry and other literatures [23]. It is obvious that reactions (4) and (5) are thermodynamically favorable, of which reaction Gibbs free energy changes are −289.0 kJ/mol and −173.0 kJ/mol at 25 °C, respectively.

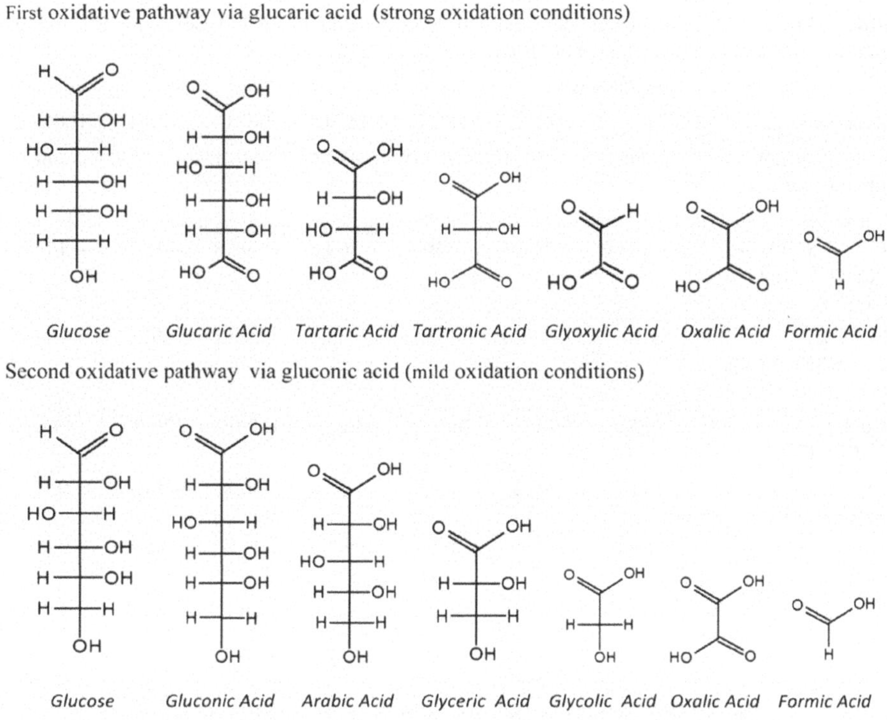

Fig. 6 Possible oxidation pathways of glucose in the H_2SO_4 leaching process [20]

However, it is well known that $LiCoO_2$ cannot react with graphite at room temperature, which means that reactions (4) and (5) need to overcome the energy barrier. And the $\Delta G(4)$ and $\Delta G(5)$ both become more negative as temperature increases, which is beneficial for the progress of reactions (4) and (5). Therefore, the reaction products will be Li_2O, CoO, Co_3O_4, CO_2. Moreover, the Gibbs free energy change of reaction (6) is below −300 kJ/mol above 500 °C, which favors that the intermediate product Co_3O_4 reacts with graphite to form product CoO and CO_2. Reaction (7) indicates that $CoCO_3$ does not form over 500 °C and reaction (8) suggests that Li_2O reacts with CO_2 to form Li_2CO_3. However, based on the variation tendency of ΔG, it can be predicted that Li_2CO_3 is unstable when temperature exceeds 900 °C. Reaction (9) concerns the possibility of forming metal cobalt and it shows that when temperature is over 600 °C, CoO can react with graphite to form Co. All reactions considered, the possible products at different temperature ranges is shown in Fig. 7. The following leaching process shows an excellent result with almost 100% leaching efficiencies of Li and Co under the optimized conditions of 2.25 M H_2SO_4, 80 °C, 30 min and S/L = 100 g/L.

Moreover, if the cathode is crushed in the pretreatment process, it is necessary to selectively remove the aluminum foil before leaching cathode material. Daniel et al. [24] proved that NaOH was very selective for the separation of Al, thus leaving all

Table 1 Possible chemical reactions of thermal decomposition of $LiCoO_2$ with graphite and the corresponding reaction heat change of Gibbs free energy [22]

	ΔG/kJ/mol						
Reactions	25	500 °C	600 °C	700 °C	800 °C	900 °C	1000 °C
$C + 12LiCoO_2 \rightarrow 6Li_2O + 4Co_3O_4 + CO_2$ (4)	−298.0	−391.4	−413.0	−434.5	−456.1	−477.7	−499.3
$C + 4LiCoO_2 \rightarrow 2Li_2O + 4CoO + CO_2$ (5)	−173.0	−311.4	−340.5	−369.6	−398.8	−427.9	−457.0
$2Co_3O_4 + C \rightarrow 6CoO + CO_2$ (6)	−181.7	−334.0	−366.0	−398.1	−430.1	−462.2	−494.2
$CoO + CO_2 \rightarrow CoCO_3$ (7)	−25.8	58.7	76.5	94.2	113.0	129.8	149.6
$Li_2O + CO_2 \rightarrow Li_2CO_3$ (8)	−175.8	−99.2	−83.1	−66.9	−50.8	−34.7	−0.1
$2CoO + C \rightarrow 2Co + CO_2$ (9)	83.8	6.8	−9.4	−25.7	−41.9	−58.1	−74.3
$CO_2 + C \rightarrow 2CO$ (10)	120.1	36.6	19.0	1.5	−16.1	−33.7	−51.3

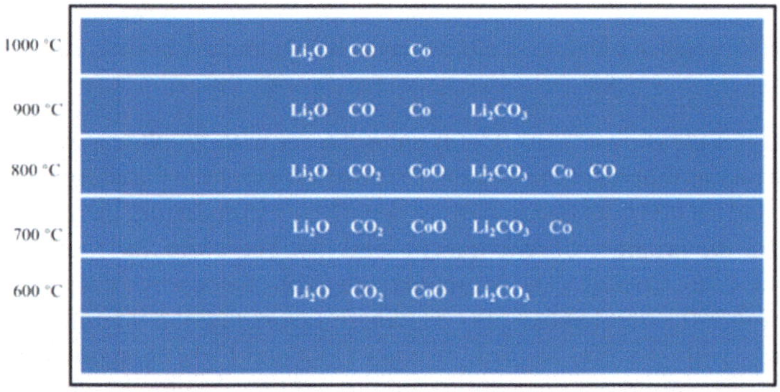

Fig. 7 Possible reaction intermediate and final products with respect to temperature [22]

Co and majority of Li in the solid phase. In the subsequent H_2SO_4 leaching, 97% Co was leached in one single stage. Chen et al. [17] also reported that after the cathode material was first leached with 5 wt% NaOH solution for the selective removal of aluminum, leaching efficiency of cobalt was 95% and lithium was 96% by H_2SO_4 and H_2O_2.

2.4 HCl Leaching

HCl has also been utilized as low costly and effective leachant in the leaching process of hydrometallurgically recycling spent lithium- ion batteries. Because of the relatively strong reductive ability, HCl alone is able to dissolve and leach cathode material. For example, during the HCl leaching process for spent LCO materials, the Co(III) can be readily reduced to Co(II) that is highly soluble in aqueous phase, the chemical reaction equation is as follows:

$$8HCl + 2LiCoO_2 \rightarrow 2CoCl_2 + 2LiCl + 5H_2O + Cl_2 \uparrow \tag{11}$$

Benefiting from the high reducing ability, the leaching efficiency of Co without the reducing additives follows the order of HCl > $H_2SO_4 \approx HNO_3$, as demonstrated by Joulie et al. [25]. Zhang et al. [26] has ever studied the reducing ability of three reagents: sulfurous acid (H_2SO_3), hydroxylamine hydrochloride (NH_2OH•HCl) and hydrochloric acid (HCl). It was found that HCl was the most suitable leachant among the three agents and leaching efficiency of more than 99% of cobalt and lithium could be achieved when 4 M HCl solution was used at a temperature of 80 °C and a reaction time of 1 h. The HCl is also more preferable for economic reasons. Li et al. [27] also reported that, a high leaching efficiency was also achieved: 97% of the Li and 99% of the Co were dissolved when the temperature was 80 °C, H^+ concentration was 4 M and reaction time was 2 h. As for leaching mixed cathode materials, Wang et al. [28] reported that a leaching efficiency of more than 99% of Co, Mn, Ni, Li could be achieved with a 4 M hydrochloric solution, 80 °C leaching temperature, 1 h leaching time and 0.02 g/mL. And in Barik et al.'s work [29], higher than 99% leaching efficiency of Co, Mn and Li was also achieved under the optimum leaching conditions: pulp density 20% (w/v), HCl 1.75 M, temperature 50 °C and time 2 h. H_2O_2 assisted HCl acid leaching was also feasible [30]. However, HCl leaching could induce Cl_2 production and HCl evaporation, which could cause environment pollution.

Some potential sources of acid can be also utilized in acid leaching sources. Polyvinyl chloride (PVC) that is widely used in the plastics industry, also face the challenges of recycling. To develop a reliable, safe and environmentally friendly method to dispose this Cl-containing polymer is urgently necessary. Recently, Liu et al. [31] used PVC as a source of HCl through the dechlorination of PVC during subcritical water oxidation, as illustrated in Fig. 8. The as-prepared HCl then was beneficial to metal leaching at the same time. This co-treatment process of both spent LIBs and PVC can achieve high leaching efficiencies of 98% for Li and 95% for Co. The novel concept of waste utilization to produce valuable materials has been also applied in the regeneration of electrode materials [32].

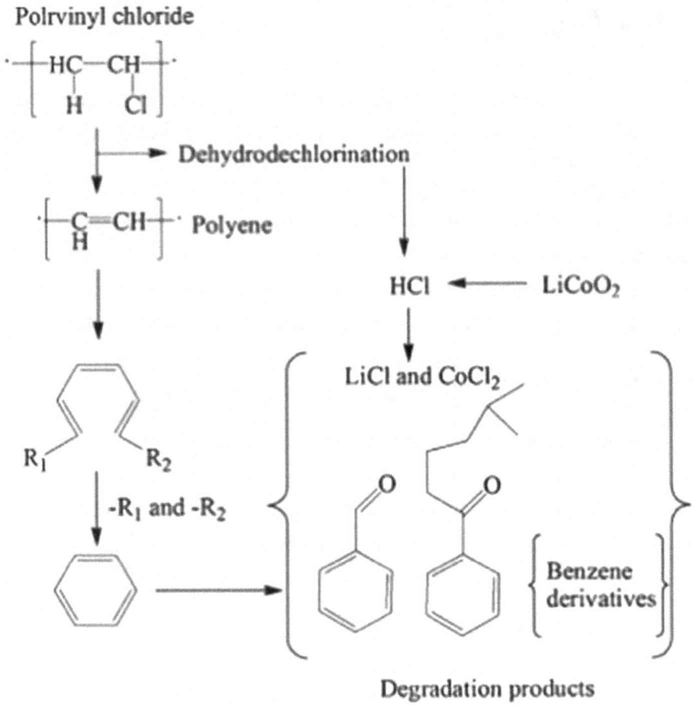

Fig. 8 Interaction route of PVC and $LiCoO_2$ during co-treatment in water [31]

2.5 *Other Inorganic Acid Leaching*

HNO_3 has also been applied to leach cathode materials. The chemical reaction equation is as follows:

$$3HNO_3 + LiCoO_2 \rightarrow LiNO_3 + Co(NO_3)_2 + 1.5H_2O + 0.25O_2 \uparrow \quad (12)$$

In order to increase the solubility of $LiCoO_2$ in HNO_3 solution, it was necessary to heat the solution [33]. And according to the results, most of the water was evaporated during heating of the HNO_3 solution. Thus, double-distilled water was continuously added to the leaching HNO_3 solution to maintain an appropriate pH value of 2.6. The solubility of Li and Co from a waste $LiCoO_2$ cathode and commercial $LiCoO_2$ in HNO_3 leaching solution is shown in Table 2. Compared with fresh $LiCoO_2$, waste

Table 2 Compositional analysis of dissolved $LiCoO_2$ in HNO_3 solution [33]

Sample	Li (ppm)	Co (ppm)
Commercial $LiCoO_2$	279.62	1387.54
Waste $LiCoO_2$	249.23	1449.90

$LiCoO_2$ produced a lower concentration of Li^+, but a higher concentration of Co^{2+}. This may be due to the changed spinel structure of $LiCoO_2$, which is the irreversibility of charge-discharge of Li^+ in the operation of the lithium ion battery. Reductive leaching cathode materials by adding H_2O_2 into HNO_3 is another measure. Lee et al. [34, 35] reported that by addition of hydrogen peroxide to the nitric acid used for leaching, the leaching efficiency increased by 45% for Co and 10% for Li compared to that in only nitric acid leaching and the most effective condition for leaching would be 1 M HNO_3, 10 to 20 g/L initial S: L ratio, 75 °C, 1.7 vol.% H_2O_2 addition and 1 h.

Compared with other inorganic acids, $\mathbf{H_3PO_4}$ has attracted the least attention due to its relatively mild acidity and weak reducibility. Chen et al. [8] assumed that the chemical process of H_3PO_4 leaching could be divided into two steps, including (A): the generation of $Co_3(PO_4)_2$ and Li_3PO_4 precipitates (Reactions (13)–(14)) and (B): the dissolving of the above two precipitates to dissolvable $Co(H_2PO_4)_2$ and LiH_2PO_4 (Reactions (15) and (16)). The leaching and separation of cobalt and lithium can be achieved simultaneously.

$$LiCoO_2(s) + 3H^+(aq) + H_2O_2(l) = Co^{2+}(aq) + Li^+(aq) + 3/2H_2O + 1/4O_2(g) \tag{13}$$

$$Co^{2+}(aq) + Li^+(aq) + PO_4^{3-}(aq) = 1/3Co_3(PO_4)_2(s) + 1/3Li_3PO_4(s) \tag{14}$$

$$Li_3PO_4(s) + 6H^+(aq) + 2PO_4^{3-}(aq) = 3Li^+(aq) + 3H_2PO_4^-(aq) \tag{15}$$

$$Co_3(PO_4)_2(s) + 12H^+(aq) + 4PO_4^{3-}(aq) = 3Co^{2+}(aq) + 6H_2PO_4^-(aq) \tag{16}$$

Based on their thermal-dynamics calculations, reactions (13)–(16) are spontaneous chemical reactions and both $Co_3(PO_4)_2$ and Li_3PO_4 can be generated firstly and then dissolved under the excessive phosphoric acid conditions. Therefore, leaching conditions have to be controlled to promote the desired reactions (reactions (13)–(15)) and prevent the undesired one (reaction (16)). Under the opitimized leaching conditions of 40 °C, 60 min, 4 vol.% H_2O_2 and 20 mL/g (L/S) and 0.7 mol/L H_3PO_4, over 99% Co could be separated and recovered directly as $Co_3(PO_4)_2$ precipitate with a purity of 97.1%. In another case, Pinna et al. [36] used phosphoric acid and hydrogen peroxide as the leaching agent and the reductant respectively and 99% dissolution efficiency of $LiCoO_2$ was achieved. But lithium and cobalt were in the form of Li^+ and Co^{2+} in the solution (Table 3).

2.6 Organic Acid Leaching

The inorganic acids shows high efficiencies for the leaching of cathode active materials, and more than 99% Co and Li can be recovered under the appropriate conditions.

Table 3 Cases of leaching by inorganic acid

Material	Leachant	Tem/°C	Time/h	S/L or pulp density (g/L)	Leaching efficiency	References
LCO	3 M H_2SO_4	70	6	1:5	Li: 98% Co: 98%	[11]
Mixed cathodes	1 M H_2SO_4	25	0.5	180	Zn: 100% Co: 70% Ni: 70% Mn: 30%	[12]
NCM	1 M H_2SO_4	95	4	50	Li: 93.4% Co: 66.2% Ni: 96.3% Mn: 50.2%	[4]
LCO	2 M H_2SO_4 + 2.0% H_2O_2 (v/v)	60	2	33	Li: 87.5% Co: 96.3%	[13]
LCO	2 M H_2SO_4 + 5% H_2O_2 (v/v)	75	1	100	Li: 99.1 Co: 70.0%	[14]
LCO	2 M H_2SO_4 + 5% H_2O_2 (v/v)	75	0.5	100	Li: 94% Co: 93%	[15]
LCO	2 M H_2SO_4 + 1 wt% glucose	80	2	20	Li: 92% Co: 88%	[20]
LCO	2.25 M H_2SO_4	80	0.5	100	Li: 100% Co: 100%	[22]
Spent LIBs	2 M H_2SO_4 + 6% H_2O_2 (v/v)	60	1	28	Li: 97% Co: 98%	[16]
mixed	4 M H_2SO_4 + 10% H_2O_2 (v/v)	85	2	10:1	Li: 96% Co: 95%	[17]
LCO	4 M HCl	80	1	1:10	Li: > 99% Co: > 99%	[26]
LCO	4 M HCl	80	2		Li: 97% Co: 99%	[27]
Mixture	4 M HCl	80	1	20	Li: >99% Co: >99% Mn: >99% Ni: >99%	[28]
Mixture	1.75 HCl	50	2	200	Li: >99% Co: >99% Mn: >99%	[29]

(continued)

Table 3 (continued)

Material	Leachant	Tem/°C	Time/h	S/L or pulp density (g/L)	Leaching efficiency	References
LCO	PVC/$LiCoO_2$ = 3: 1	350	0.5	16	Li: 98% Co: 95%	[31]
LCO	1 M HNO_3 + 1.7% H_2O_2 (v/v)	75	0.5	20	Li: >95% Co: >95%	[35]
LCO	2% H_3PO_4 + 2% H_2O_2 (v/v)	90	1	8	Li: 88% Co: 99%	[36]

However, some disadvantages limits the practical application of inorganic acids. First, some hazardous gases such as Cl_2, SO_3 and NO_x can be inevitably produced in the leaching process, which is a threat to environment and human health [37]. Second, the leachants with low pH cannot be recovered directly in the following procedure and make the recovery technology more complicated. For example, if precipitation method is employed to remove some impurities in the leachates such as Al, Cu and Fe, or recover the valuable metals such as Co, Ni and Mn, more alkaline is needed to neutralize the residual acid. Third, the waste water containing strong acid from the leaching process must be post-treated before emission to avoid the secondary pollution. Thus, a further disposal of the hazardous gases, acidic leachants and acid waste water is required and additional expenditure will be produced by using inorganic acids as leaching reagents [38].

Because of the properties of easy degradation and little environment pollution, organic acids have been increasingly important for environmentally friendly recycling process in recent years. The organic acids can replace the typically used acids without sacrificing the leaching efficiency. Moreover, they usually possess chelating or complexing properties, which offers possibilities for the subsequent recycling process. Though the specific reactions of different organic acids may be various, the overall chemical reaction can be summarized as below:

$$6H^+ + 2LiCoO_2 + H_2O_2 \rightarrow 2Li^+ + 2Co^{2+} + 4H_2O + O_2 \uparrow \quad (17)$$

Li et al. [39] reported a series of studies of leaching cathode materials by different kinds of organic acids. It was found that DL-malic acid, ascorbic acid, citric acid, succinic acid and lactic acid all could be used as leachants for waste cathode materials. Considering both leaching efficiency and chemical consumption, nearly 100% Li and more than 90% Co could be leached under the conditions of 1.5 M DL-malic acid, 2.0 vol.% hydrogen peroxide, a leaching temperature of 90 °C, a S: L ratio of 20 g/L and a time interval of 40 min. 1.5 mol/L Succinic acid with 4 vol.% H_2O_2 was also investigated to leach LCO and more than 96% Li and nearly 100% Co were leached at 70 °C, 15 g/L and for 40 min. Under the conditions of 90 °C, 20 g/L and 30–40 min, recoveries of nearly 100% of Li and in excess of 90% Co were achieved following leaching in citric acids. With the assistant of H_2O_2, lactic acid was able to dissolve the collected and pretreated cathode material and the leaching efficiencies of Li, Ni, Co, Mn reached 99.7%, 98.2% 98.9% and 98.4% respectively when the optimum conditions were 1.5 mol/L lactic acid concentration, 20 g/L, 70, 0.5 vol.% H_2O_2 content and reaction time of 20 min. Due to the reducibility of ascorbic acid, H_2O_2 was unnecessarily added in the leaching process. The optimized leaching conditions were explored to be 1.25 mol/L ascorbic acid concentration, leaching temperature of 70 °C, leaching time of 20 min and S/L ratio of 25 g/L and as much as 94.8% Co and 98.5% Li can be recovered. In addition, formic acid has been proved to be effective to dissolve cathode material with H_2O_2 as reductant by Gao et al. [40]. L-Tartaric acid with H_2O_2 was also applied as leachant with the leaching efficiencies of 99.07% for Li, 98.64% for Co, 99.31% for Mn and Ni under the optimized conditions [40]. Replacing H_2O_2 by ascorbic acid as reductant in tartaric acid leaching solution was also feasible [41]. Another case of using organic acid to replace H_2O_2 as reductant was performed by Chen et al. [42]. They investigated the leaching effect of citric acid (H_3Cit) but with different reductants: Tea Waste (TW), Phytolacca Americana (PA), and H_2O_2. It was found that at the optimal conditions, H_3Cit and TW system reached leaching efficiencies of 96% Co and 98% Li, which was similar to 98% Co and 99% Li in H_3Cit and H_2O_2 system. For H_3Cit and PA system, inferior leaching results (83% Co and 96% Li) could be obtained under the optimized conditions (Table 4).

2.7 *Alkaline Leaching*

Compared with acid leaching, leaching cathode materials in alkaline solution has rarely been reported. However, Alkali leaching with an ammonia-based system is relatively selective for specific elements, such as Ni, Co, and Li, because of the formation of stable metal ammonia complexes. It has been proved to be feasible for leaching Ni and Co. The reactions are as follows:

$$Ni^{2+} + nNH_3 \rightarrow Ni(NH_3)n^{2+} \quad (18)$$

$$Co^{2+} + nNH_3 \rightarrow Co(NH_3)n^{2+} \quad (19)$$

According to the Eh–pH diagrams of the Co–NH_3–H_2O system and Ni–NH_3–H_2O system, $Co(NH_3)_6^{3+}$, $Co(NH_3)_5^{2+}$ and $Co(NH3)_4^{2+}$ are the major soluble species in the solution over the pH range of 9–11, and $Ni(NH_3)_6^{3+}$ is the predominant species in the range of pH from 8.5 to 10.5 [48]. Ku et al. [49] designed an alkali leaching system, in which NH_3 and $(NH_4)_2CO_3$ act as the leaching reagents and ammonium sulfite [$(NH_4)_2SO_3$] act as the reducing agent to investigate the leaching behavior of Ni, Co, Mn, Al, and Cu. It was found that Co and Cu were leached out completely as they can form stable $Co(NH_3)_6^{2+}$ and $Cu(NH_3)_4^{2+}$ complex ions. Ni leached out

Table 4 Cases of leaching by organic acid

Material	Leachant	Tem/°C	Time/h	S/L or pulp density/g/L	Leaching efficiency	References
LCO	1.5 M DL-malic acid +2% H_2O_2(v/v)	90	0.67	20	Li: 100% Co: >90%	[39]
LCO	1.25 M Ascorbic acid	70	0.22	25	Li: 98.5% Co: 94.8	[43]
LCO	1.5 M Succinic acid +4% H_2O_2(v/v)	70	0.67	15	Li > 96% Co: 100	[44]
LCO	Citric acid + H_2O_2	90	0.67	20	Li: 100% Co: >90%	[45]
Mixture	1.5 M Lactic acid +0.5% H_2O_2(v/v)	70	0.33	20	Li: 97.7% Co: 98.9% Mn: 98.4% Ni: 98.2%	[42]
LCO	2 M H_3Cit + 0.6 g/g H_2O_2	70	1.33	50	Li: 99% Co: 98%	[46]
NCM	2 M Formic acid +6% H_2O_2(v/v)	60	0.5	50	Li: 99.93%	[40]
Mixture	2 M L-Tartaric + 4% H_2O_2(v/v)	70	0.5	17	Li: 99.07% Co: 98.64% Mn, Ni: 99.31%	[47]

moderately, while Mn and Al hardly leached out at all. The unstable nature of Mn and Al in ammonia-based solution resulted in the formation of the corresponding Al_2O_3, $MnCO_3$, and Mn oxides, as revealed in the XRD pattern of the leaching residue. Zheng et al. [50] reported that 95.3% Li, 80.7% Co, 89.8% Ni and only 1.3% Mn could be leached out under the optimum conditions. In this alkali leaching system, NH_3 and $(NH_4)_2SO_4$ were employed as the leaching solution and sodium sulfite Na_2SO_3 as the reducing agent. $(NH_4)_2MnSO_3 \bullet H_2O$ was found in the residue, which was explained by the reduction of Mn^{4+} to Mn^{2+} and subsequent precipitation. Similar result was attained by Wang et al. [51] (Figs. 9 and 10).

3 Purification and Recovery

After leaching, one of the main aim to recovery spent materials is to obtain pure metal or chemical compounds. How to separate and recover the different metal ions in the leaching solution is the next key issue. Several methods such as solvent extraction,

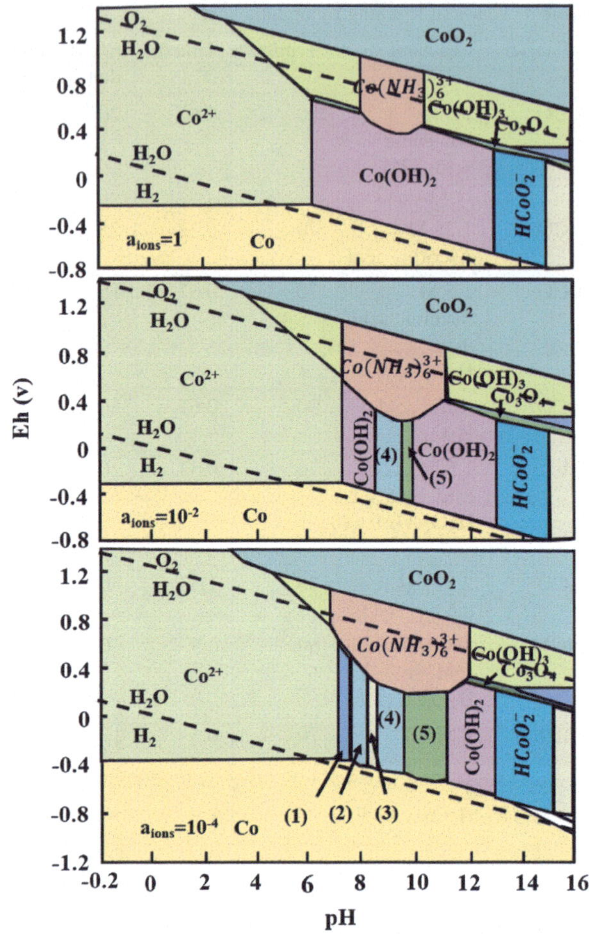

Fig. 9 Eh–Ph diagram for the Co–NH3–H2O system at 25 °C and 101.3 kPa, Activity of total dissolved ammonia species = 1. 1, $Co(NH_3)^{2+}$; 2, $Co(NH_3)_2^{2+}$; 3, $Co(NH_3)_3^{2+}$; 4, $Co(NH_3)_4^{2+}$; 5, $Co(NH_3)_5^{2+}$ [50]

chemical precipitation and electro-deposition (electro-winning) have been proposed and reported massively. Before extraction and chemical precipitation, the Li ion concentration in leaching solution could be enriched by reverse osmosis strategy as clarified by Swain [52]. However, the actual situation in practical production is that the metal ions in leaching liquor are usually substantial and not necessarily enriched. And recently, the novel regeneration method to recycle the spent materials has become the new direction, which refers to resynthesizing the cathode material directly from the leaching liquor. This seems a little more sophisticated and energy-effective since the intermediate processes of raw chemical material and cathode material synthesis are integrated.

Fig. 10 Eh–Ph diagram for the Ni–NH_3–H_2O system at 25 °C and 101.3 kPa, Activity of total dissolved ammonia species = 1. 1, $Ni(NH_3)^{2+}$; 2, $Ni(NH_3)_2^{2+}$; 3, $Ni(NH_3)_3^{2+}$; 4, $Ni(NH_3)_4^{2+}$; 5, $Ni(NH_3)_5^{2+}$; 6,$Ni(NH_3)_6^{2+}$ [50]

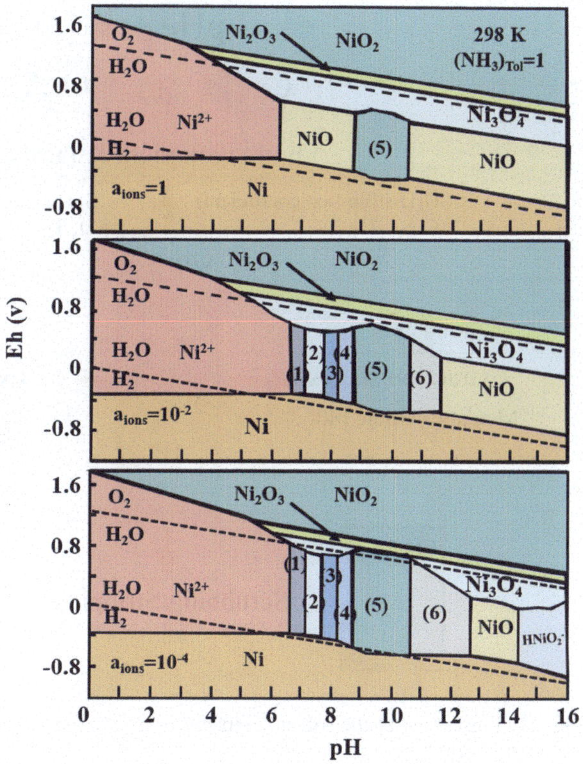

3.1 Solvent Extraction

Solvent extraction is based on the different solubility of compounds using a two-liquid phase system, normally organic and aqueous phases. When the extraction system reaches equilibrium, each liquid contains different concentration of metal ions so that separation could be realized as shown in Fig. 11. Stripping the loaded organic with some solution like H_2SO_4, the extractant can be reused. The effectiveness of solvent extraction can be determined by the extraction yield and phase separation ability. The distribution ratio (D) that is the mass ratio of certain metal ions in the organic phase to that in the aqueous phase at equilibrium. The larger D means a higher extraction yield of metal ions [53]. The $pH_{1/2}$ value and separation factor (SF) are employed to define the separation ability of different metal ions. The $pH_{1/2}$ value is the equilibrium pH value when 50% of the metal ions are extracted, while the SF value between A and B is calculated based on the equation $SF = D_A/D_B$ [54]. Thus, the larger value of $pH_{1/2}$ and SF indicate a separation performance.

Extraction method has been widely used in various hydrometallurgy processes and proved to be an efficient method for the separation of valuable metal ions or the removal of impurities from the aqueous leaching solution. Up to now, several

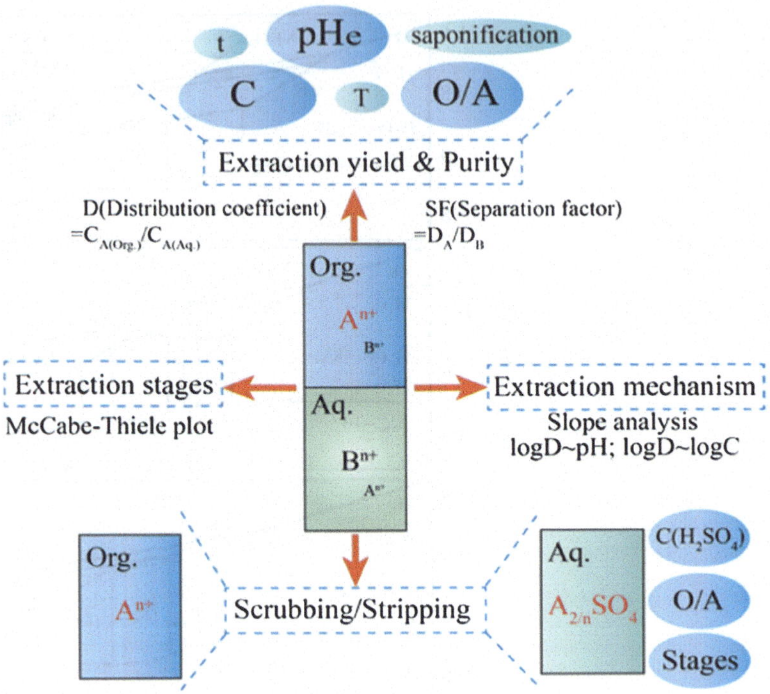

Fig. 11 Illustration of the solvent extraction process, mechanism analysis, and factors related to extraction, scrubbing, and stripping [55]

solvents such as Cyanex 272 (bis-2,4,4-trimethylpentyl phosphinic acid), D2EHPA (di(2-ethylhexyl) phosphoric acid), PC88A (2-ethylhexyl phosphinic acid mono-2-ethylhexyl ester) and so on have been widely utilized to separate metal ions in the spent lithium-ion battery leaching solution.

Solvent extraction with PC88A is very effective in separating cobalt from other metals in the leach liquor. After extraction with 0.90 M PC-88A in kerosene at an O:A ratio of 0.85:1 and pH = 6.7 in a single stage, which was followed by single stage lithium scrubbing from the loaded solvent by a chloride solution containing 30 g/L of cobalt at an initial pH of 1.0 and an O:A ratio of 10:1 and then by stripping with a 2 M H_2SO_4 solution at an O:A ratio of 5:1, the purity of the cobalt recovered could reach 99.99% or better [26]. Yang et al. reported that almost 100% cobalt could be extracted from leaching liquor when the solvent extraction was conducted at 25 °C and pH = 5.5 with 35% PC88A, A:O = 0.5 [22].

Cyanex 272 is another very effective solvent that can selectively separate cobalt from other transition metal ions in the leaching liquor. Kang et al. [16] found that cobalt could be selectively extracted from the purified aqueous phase by equilibrating with 50% saponified 0.4 M Cyanex 272 at an equilibrium pH ~ 6. The McCabe–Thiele plot predicted that 99.9% cobalt extraction in a 2 stage counter-current operation with A/O ratio of 1/2. Separation factors for the extraction of

Co/Li and Co/Ni at pH 6 were close to 750. The stripping of the loaded organic phase with 2 M H_2SO_4 produced a solution of 96 g/L Co from which pure pigment grade cobalt sulfate could be recovered by evaporation/recrystallisation. By adjusting the adding amount of KOH, the separation factor between Co and Ni could reach 1180.37 [56]. Wang et al. [57] investigated the extraction and separation of cobalt(II), copper(II) and manganese(II) by Cyanex272, PC88A and their mixtures and found that the mixed extractants system had an evident synergistic effect for Mn, Co and Cu, especially at higher pH_e and lower concentration of Na_2SO_4. The maximum synergistic enhancement co-efficiency, Rmax, were obtained at the mole fraction of $X_{Cyanex272}$ around 0.5. The order of synergistic enhancement effect was Mn > Co > Cu.

D2EHPA has been studied to separate manganese from other metals in leaching liquor. Chen et al. [58] reported that extraction percentage of manganese as high as 97.1% by Co-D2EHPA could be attained under the optimized conditions: extraction time 5 min, equilibrium pH 3.5, 15 vol.% Co-D2EHPA and O:A of 1:1.In Wang et al's work [59], the extraction of cobalt metal ion was based on the first step leaching experiment, using D2EHPA to extract copper and manganese. Under the conditions that D2EHPA saponification rate was 20%, the sulfonated kerosene accounted for the volume ratio of 70%, O/A ratio of 1:1, manganese was separated and the purity of cobalt in final product was up to 99.5%.

Cerpa et al. [60] used DP-8R (di(2-ethylhexyl) phosphoric acid) and Acorga M5640 (hydroxyoxime) diluted in Exxsol D100 to co-extract cobalt and nickel from aqueous acidic sulfate media. Selective cobalt stripping from the loaded organic solution was achieved in one stage with 12 g dm^{-3} H_2SO_4, 1 min equilibration time and 20 °C. The cobalt-nickel separation factor was near 270. Thus, the objective of separating the two metals through co-extraction and selective stripping was accomplished. However, Suzuki et al. [61] found that although Acorga M5640 provided higher cobalt selectivity, its stripping efficiency of cobalt was low (less than 10%). So based on the experiment, they concluded that PC88A/TOA(tri-n-octylamine) mixed extractant was suitable for the separation of cobalt and lithium. While a few literature reported that Acorga M5640 and Cyanex 272 could be used in combination to effectively purify the metal ions [11, 54].

The experimental design needs to be optimized to find the best extraction conditions. The extraction process is usually performed at room temperature and can be finished within 30 min. These are both advantages of the solvent extraction method. Meanwhile, the pH of all the extraction experiments indicate that an acid solution is found at the equilibrium state. In Fig. 12, the suitable scales of pH of different extraction reagents are summarized. It is obvious that the suitable pH value to separate Co and Ni by some extract solvents lies in the pH range between 3 and 5, which requires a good performance in corrosion resistance for the reactor [2, 62]. After the extraction process there are two more necessary steps under normal conditions. One is the scrubbing step to remove the co-extracted impurities and the other is stripping to transfer the metal ions from the organic phase to aqueous solution for final treatment. H_2SO_4 is the most commonly used stripping agent.

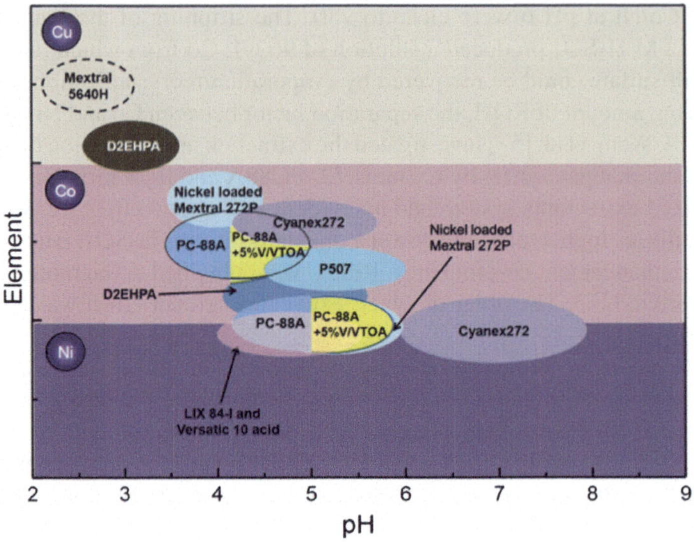

Fig. 12 Effect of suitable pH on the solvent extraction of different extraction reagents (conditions: $T = 25$ °C and O:A = 1:1) [2]

3.2 *Chemical Precipitation and Electrodeposition*

Chemical precipitation and electro-deposition (electro-winning) are two main ways to recycle the metals of interest in the form of raw materials after leaching and extraction.

Electro-deposition is strategic to induce redox reactions between the ions in the solution by applying a voltage between two electrodes. Electrodeposition has the advantage of achieving a high product purity because it does not induce other impurities, but it has a high energy consumption because of the use of electricity. Myoung et al. [33] applied a linear sweep voltammogram to dissolved $LiCoO_2$ in HNO_3 solution for a Ti cathode. The curve indicated that a multi-step electroreduction occurred. The detailed reaction mechanisms are considered to be (Fig. 13):

$$2H_2O + O_2 + 4e^- \rightarrow 4OH^- \tag{20}$$

$$NO^{3-} + H_2O + 2e^- \rightarrow NO_2 + 2OH^- \tag{21}$$

$$Co^{3+} + e^- \rightarrow Co^{2+} \tag{22}$$

$$Co^{2+} + 2OH^-_{ad}/Ti \rightarrow Co(OH)_2/Ti \tag{23}$$

Fig. 13 Linear sweep voltammograms for different concentrations of waste $LiCoO_2$ in 0.1 M HNO_3 sulution on a Ti electrode. **a** 10 mM $LiCoO_2$; **b** 100 mM $LiCoO_2$. Scan rate = 10 mV/s. Bulk pH = 2.59 [33]

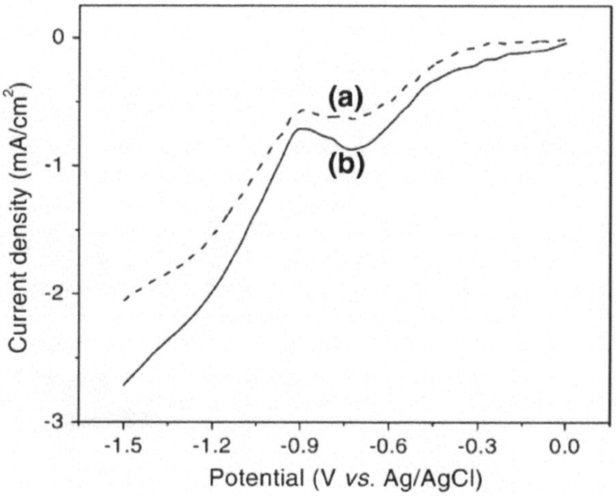

In the electrodeposition, pH is a very important factor that can heavily influence charge efficiency. Adding H_3BO_3 to the solution could limit pH variations in the interface electrode solution, which is favorable for high current efficiencies [30, 63, 64]. Freitas et al. [30] found that the largest charge efficiency was 96.9% at pH 5.4 at 10.0 C cm^{-2} charge density. The application of the nucleation models to the initial electrodeposition stages shows that at pH 5.4, the nuclei grow progressively. By the way, the cobalt oxide product from electrodeposition could also be used as supercapacitor material [65, 66].

After separation, chemical precipitation is another method to recover the metals of interest as raw materials. The separation mechanism of chemical precipitation relies on the different solubilities of metal compounds at certain pH. Generally, transition metal hydroxides and oxalates have much lower solubility compared with that of the corresponding lithium compounds. Meanwhile, impurity metal ions, like Fe^{3+}, Al^{3+}, and Cu^{2+}, usually precipitate at a relatively lower pH (Fig. 14) [27, 67]. For lithium to be precipitated, the two commonly used chemical agents are Na_2CO_3 [13, 15, 40] and Na_3PO_4 [27, 36, 46] and high purity lithium salts could be obtained. Gao et al. [40] obtained 99.9% purity Li_2CO_3 and Pinna et al. [36] precipitated Li_3PO_4 with a purity of 98.3%. For transition metals to be precipitated, $H_2C_2O_4$ is widely used [13, 36, 46] and NaOH is also reported to precipitate manganese [68].

In addition, the very intriguing cases by selective leaching-precipitation and separation are worth mentioning. One is that Cai et al. [70] according to the phase diaphragm added Na_2S to the $LiCo_{0.5}Mn_{0.5}O_2$ leaching solution dissolved in H_2SO_4 and H_2O_2 to precipitate out a mixture of CoS and MnS. CoS and MnS are further separated by adding acetic acid where MnS is selectively dissolved while CoS remains in the solid phase. After filtering out the undissolved CoS, MnS or $Mn(OH)_2$ is precipitated by adding NaOH. The other case is that Katsiapi et al. [71] adopted

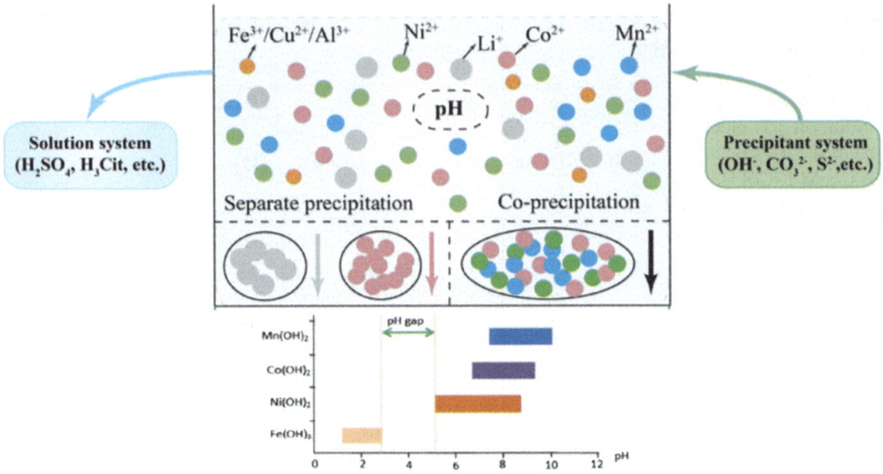

Fig. 14 pH values of the start and end of hydroxide precipitation for different metal ions [69, 55]

ammonia-ammonium carbonate solution to selectively leach cobalt from the as precipitated Co–Mn hydroxide and under optimum conditions, cobalt recovery reached 93% whereas manganese leaching did not exceeded 0.05%.

3.3 *Regeneration*

Regeneration means that cathode material is synthesized directly from spent lithium-ion battery leaching liquor. Currently, there are two main ways to regenerate cathode materials: one is sol-gel and the other is co-precipitation. Before regeneration, the contents of each metal ions must be analyzed accurately and then adjusted to the desirable value by adding certain chemical compounds.

The organic acids leaching solution are usually disposed by sol-gel method since the existing organic ingredients can serve as chelates. After adding acetic salts or nitrate to adjust the metal ions ratio, sol is obtained by increasing pH of the solution and gel is obtained by drying gel and cathode material will be synthesized by calcinating the gel. By doing so, Li et al. [72] reported that the discharge capacity of re-synthesized $LiCoO_2$ was 137 mAh/g and the capacity retention were 97.98 and 88.14% after 20 cycles and 40 cycles at 0.1 C. And the regenerated $LiNi_{1/3}Co_{1/3}Mn_{1/3}O_2$ cathode material delivered a highly discharge capacity, 138.2 mAh/g at 0.5 C with a capacity retention of 96% after 100 cycles [42]. In Yao et al. work [73, 74], the capacity of re-synthesized $LiNi_{1/3}Co_{1/3}Mn_{1/3}O_2$ could reach 147 mAh/g. However, nitrate was applied in adjusting metal ions ratio process, which would produce nitrogen and bring about environmental pollution in gel calcination process.

In co-precipitation cases, sulfate is utilized to adjust the ratio of metal ions when the spent cathode material is leached by sulphuric acid [70, 75]. By adding NaOH or ammonia to increase pH of the adjusted liquor, transition metal hydroxide precursor could be co-precipitated. Then by calcinating the stoichiometric mixture of Li_2CO_3 and the precursor, re-synthesized cathode material is obtained. Sa et al. [75, 76] reported that all the resynthesized NCM cathode materials delivered more than 150 mAh/g discharge capacity at 0.1C. Zou et al. [69] resynthesized $LiNi_{0.33}Mn_{0.33}Co_{0.33}O_2$ and its initial capacity was 130.2 mAh/g (voltage range of 2.5–4.6 V at 46.6 mA/g) and the reversible capacity after 50 cycles was 107.29 mAh/g. In yang et al's work [77], the initial discharge capacities (0.1 C, 2.7–4.3 V) of the regenerated $LiNi_{0.8}Co_{0.1}Mn_{0.1}O_2$, $LiNi_{0.5}Co_{0.2}Mn_{0.3}O_2$, $LiNi_{0.33}Co_{0.33}Mn_{0.33}O_2$ were 197.7 mAh/g, 174.3 mAh/g, 168.3 mAh/g and the capacity retention were 86.3, 95 and 96% after 50 cycles respectively. Additionally, Zhang et al. [78] proposed a novel application of oxalic acid leaching to regenerate $LiNi_{1/3}Co_{1/3}Mn_{1/3}O_2$ cathodes from spent LIBs. With lithium dissolving into the solution, the transition metals transformed into oxalate precipitates and deposited, thus separating lithium and transition metals in one simple step. After mixing with certain amount of Li_2CO_3, the oxalate precipitated together with unreacted NCM are directly calcined into new NCM cathode. The regenerated NCM delivered the initial discharge capacity of 168 mAh/g at 0.2C and 153.7 mAh/g after 150 cycles with a high capacity retention of 91.5%. Another route of co-precipitation is that the metal ion ratio is adjusted by acetic salts and all the metals including lithium are co-precipitated by oxalate acid. Through this way, Li et al. [79] resynthesized $Li_{1.2}Co_{0.13}Ni_{0.13}Mn_{0.54}O_2$ with the initial capacity of 258.8 mAh/g and capacity retention of 87% after 50 cycles.

References

1. X. Zeng, J. Li, N. Singh, Recycling of spent lithium-ion battery: a critical review. Crit. Rev. Environ. Sci. Technol. **40**(10), 1129–1165 (2014)
2. W. Lv, Z. Wang, H. Cao, Y. Sun, Y. Zhang, Z. Sun, A critical review and analysis on the recycling of spent lithium-ion batteries. ACS Sustainable Chem. Eng. **6**, 1504–1521 (2018)
3. E. Billy, M. Joulié, R. Laucournet, A. Boulineau, E. De Vito, D. Meyer, Dissolution mechanisms of $LiNi_{1/3}Mn_{1/3}Co_{1/3}O_2$ positive electrode material from lithium-ion batteries in acid solution. ACS Appl. Mat. Interfaces **10**, 16424–16435 (2018)
4. P. Meshram, B.D. Pandey, T.R. Mankhand, Recovery of valuable metals from cathodic active material of spent lithium ion batteries: leaching and kinetic aspects. Waste Manage. **45**, 306–313 (2015)
5. P. Meshram, B.D. Pandey, T.R. Mankhand, Hydrometallurgical processing of spent lithium ion batteries (LIBs) in the presence of a reducing agent with emphasis on kinetics of leaching. Chem. Eng. J. **281**, 418–427 (2015)
6. G.P. Nayaka, K.V. Pai, J. Manjanna, S.J. Keny, Use of mild organic acid reagents to recover the Co and Li from spent Li-ion batteries. Waste Manage. **51**, 234–238 (2016)
7. C. Sun, L. Xu, X. Chen, T. Qiu, T. Zhou, Sustainable recovery of valuable metals from spent lithium-ion batteries using DL-malic acid: Leaching and kinetics aspect. Waste Manage. Res. **36**, 113–120 (2018)

8. X. Chen, H. Ma, C. Luo, T. Zhou, Recovery of valuable metals from waste cathode materials of spent lithium-ion batteries using mild phosphori acid. J. Hazardous Mat. **326**, 77–86 (2017)
9. X. Zhang, H. Cao, Y. Xie, P. Ning, H. An, H. You, F. Nawaz, Closed-loop process for recycling $LiNi_{1/3}Co_{1/3}Mn_{1/3}O_2$ from the cathode scraps of lithium-ion batteries: process optimization and kinetics analysis. Sep. Purif. Technol. **150**, 186–195 (2015)
10. L. He, S. Sun, X. Song, J. Yu, Leaching process for recovering valuable metals from the $LiNi_{1/3}Co_{1/3}Mn_{1/3}O_2$ cathode of lithium-ion batteries. Waste Manage. **64**, 171–181 (2017)
11. J. Nan, D. Han, X. Zuo, Recovery of metal values from spent lithium-ion batteries with chemical deposition and solvent extraction. J. Power Sources **152**, 278–284 (2005)
12. K. Tanong, L. Coudert, G. Mercier, J.-F. Blais, Recovery of metals from a mixture of various spent batteries by a hydrometallurgical process. J. Environ. Manage. **181**, 95–107 (2016)
13. S.G. Zhu, W.Z. He, G.M. Li, X. Zhou, X.J. Zhang, J.W. Huang, Recovery of Co and Li from spent lithium-ion batteires by combination method of acid leaching and chemical precipitation. Trans. Nonferrous Met. Soc. China **22**, 2274–2281 (2012)
14. M.K. Jha, A. Kumari, A.K. Jha, V. Kumar, J. Hait, B.D. Pandey, Recovery of lithium and cobalt from waste lithium ion batteries of mobile phone. Waste Manage. **33**, 1890–1897 (2013)
15. B. Swain, J. Jeong, J.-C. Lee, G.-H. Lee, J.-S. Sohn, Hudrometallurgical process for recovery of cobalt from waste cathodic active material generated during manufacturing of lithium ion batteries. J. Power Sources **167**, 536–544 (2007)
16. J. Kang, G. Senanayake, J. Sohn, S.M. Shin, Recovery of cobalt sulfate from spent lithium ion batteries by reductive leaching and solvent extraction with Cyanex 272. Hydrometallurgy **100**, 168–171 (2010)
17. L. Chen, X. Tang, Y. Zhang, L. Li, Z. Zeng, Y. Zhang, Process for the recovery of cobalt oxalate from spent lithium-ion batteries. Hydrometallurgy **108**, 80–86 (2011)
18. H. Li, S. Xing, Y. Liu, F. Li, H. Guo, G. Kuang, Recovery of lithium, iron, and phosphorus from spent $LiFePO_4$ batteries using stoichiometric sulfuric acid leaching system. ACS Sustain. Chem. Eng. **5**, 8017–8024 (2017)
19. Q. Meng, Y. Zhang, P. Dong, Use of glucose as reductant to recover Co from spent lithium ions batteries. Waste Manag. **64**, 214–218 (2017)
20. E. Pagnanelli, G. Moscardini, S. Granata, L. Cerbelli, A. Agosta, L.Toro Fieramosca, Acid reducing leaching of cathodic powder from spent lithium ion batteries: glucose oxidative pathways and particle area evolution. J. Ind. Eng. Chem. **20**, 3201–3207 (2014)
21. M. Joulie, E. Billy, R. Laucournet, D. Meyer, Current collectors as reducing agent to dissolve active materials of positive electrodes from Li-ion battery wastes. Hydrometallurgy **169**, 426–432 (2017)
22. Y. Yue, S. Wei, Y. Bu, Z. Chenyang, S. Shaole, Y. Hu, Recovering metal values from spent lithium ion battery via a combination of reduction thermal treatment and facile acid leaching. ACS Sustainable Chem. Eng. **6**(8), 10445–10453 (2018)
23. O. Kubaschewshi, C.B. Alcock, *Metallurgical Thermochemistry*, 5th edn. (Pergamon Press, UK, 1979)
24. D.A. Ferreira, L.M.Z. Prados, D. Majuste, M.B. Mansur, Hydrometallurgical separation of aluminum, cobalt, copper and lithium from spent Li-ion batteries. J. Power Sources **187**, 238–246 (2009)
25. M. Joulie, R. Laucournet, E. Billy, Hydrometallurgical process for the recovery of high value metals from spent lithium nickel cobalt aluminum oxide based lithium-ion batteries. J. Power Sources **247**, 551–555 (2014)
26. P. Zhang, T. Yokoyama, O. Itabashi, T.M. Suzuki, K. Inoue, Hydrometallurgical process for recovery of metal values from spent lithium-ion secondary batteries. Hydrometallurgy **47**, 259–271 (1998)
27. J. Li, P. Shi, Z. Wang, Y. Chen, C.C. Chang, A combined recovery process of metals in spent lithium-ion batteries. Chemosphere **77**, 1132–1136 (2009)
28. R.C. Wang, Y.C. Lin, S.H. Wu, A novel recovery process of metal values from the cathode active materials of the lithium-ion secondary batteries. Hydrometallurgy **99**, 194–201 (2009)

29. S.P. Barik, G. Prabaharan, L. Kumar, Leaching and separation of Co and Mn from electrode materials of spent lithium-ion batteries using hydrochloric acid: Laboratory and pilot scale study. J. Cleaner Prod. **147**, 37–43 (2017)
30. M.B.J.G. Freitas, E.M. Garcia, Electrochemical recycling of cobalt from cathodes of spent lithium-ion batteries. J. Power Sources **171**, 953–959 (2007)
31. K. Liu, F. Zhang, Innovative leaching of cobalt and lithium from spent lithium-ion batteries and simultaneous dechlorination of polyvinyl chloride in subcritical water. J. Hazard. Mat. **316**, 19–25 (2016)
32. X. Meng, H. Cao, J. Hao, P. Ning, Xu Gaojie, Zhi Sun, Sustainable preparation of $LiNi_{1/3}Co_{1/3}Mn_{1/3}O_2$–$V_2O_5$ cathode materials by recycling waste materials of spent lithium-ion battery and vanadium-bearing slag. ACS Sustainable Chem. Eng. **6**, 5797–5805 (2018)
33. J. Myoung, Y. Jung, J. Lee, Y. Tak, Cobalt oxide preparation from waste $LiCoO_2$ by electrochemical-hydrothermal method. J. Power Sources **112**, 639–642 (2002)
34. C.K. Lee, K.I. Rhee, Preparation of $LiCoO_2$ from spent lithium-ion batteries. J. Power Sources **109**(1), 17–21 (2002)
35. C.K. Lee, K.I. Rhee, Reductive leaching of cathodic active materials from lithium ion battery wastes. Hydrometallurgy **68**, 5–10 (2003)
36. E.G. Pinna, M.C. Ruiz, M.W. Ojeda, M.H. Rodriguez, Cathodes of spent Li-ion batteries: Dissolution with phosphoric acid and recovery of lithium and cobalt from leach liquors. Hydrometallurgy **167**, 66–71 (2017)
37. L. Sun, K. Qiu, Organic oxalate as leachant and precipitant for the recovery of valuable metals from spent lithium-ion batteries. Waste Manage. **32**, 1575–1582 (2012)
38. Y. Yao, M. Zhu, Z. Zhao, B. Tong, Y. Fan, Z. Hua, Hydrometallurgical processes for recycling spent lithium-ion batteries: a critical review. ACS Sustainable Chem. Eng. **6**(11), 13611–13627 (2018)
39. L. Li, J. Ge, R. Chen, F. Wu, S. Chen, X. Zhang, Environmental friendly leaching reagent for cobalt and lithium recovery from spent lithium-ion batteries. Waste Manage. **30**, 2615–2621 (2010)
40. W. Gao, X. Zhang, X. Zheng, X. Lin, H. Cao, Y. Zhang, Z. Sun, Lithium carbonate recovery from cathode scrap of spent lithium-ion battery-a closed-loop process. Environ. Sci. Technol. **51**(3), 1662–1669 (2017)
41. G.P. Nayaka, K.V. Pai, G. Santhosh, J. Manjanna, Dissolution of cathode active material of spent Li-ion batteries using tartaric acid and ascorbic acid mixture to recover Co. Hydrometallurgy **161**, 54–57 (2016)
42. L. Li, E. Fan, Y. Guan, X. Zhang, Q. Xue, L. Wei, F. Wu, R. Chen, Sustainable recovery of cathode materials from spent lithium-ion batteries using lactic acid leaching system. ACS Sustainable Chem. Eng. **5**(6), 5224–5233 (2017)
43. L. Li, J. Lu, Y. Ren, X. Zhang, R. Chen, F. Wu, K. Amine, Ascorbic-acid-assisted recovery of cobalt and lithium from spent Li-ion batteries. J. Power Sources **218**, 21–27 (2012)
44. Li Li, Qu Wenjie, Xiaoxiao Zhang, Lu Jun, Renjie Chen, Wu Feng, Khalil Amine, Succinic acid-based leaching system: a sustainable process for recovery of valuable metals from spent Li-ion batteries. J. Power Sources **282**, 544–551 (2015)
45. L. Li, J.B. Dunn, X. Zhang, L. Gaines, R.J. Chen, F. Wu, K. Amine, J. Power Sources **233**, 180–189 (2013)
46. X. Chen, C. Luo, J. Zhang, J. Kong, T. Zhou, Sustainable recovery of metals from spent lithium-ion batteries: a green process. ACS Chem. Eng. **3**(12), 3104–3113 (2015)
47. P. Li, Y. Shu, Y. Yan, F. Xing, G. Jian, Recovery of lithium Nickel, cobalt, and manganese from spent lithium-ion batteries using L-Tartaric acid as leachant. ACS Sustainable Chem. Eng. **5**(1), 714–721 (2017)
48. X. Meng, K.N. Han, The principles and applications of ammonia leaching of metals-a review. Miner. Process. Extr. Metall. Rev. **1**(16), 23–61 (1996)
49. H. Ku, Y. Jung, M. Jo, S. Park, S. Kim, D. Yang, K. Rhee, E. An, J. Sohn, K. Kwon, Recycling of spent lithium-ion battery cathode materials by ammoniacal leaching. J. Hazard. Mat. **313**, 138–146 (2016)

50. X. Zheng, W. Gao, X. Zhang, M. He, X. Lin, H. Cao, Y. Zhang, Z. Sun, Spent lithium-ion battery recycling –Reductive ammonia leaching of metals from cathode scrap by sodium sulphite. Waste Manage. **60**, 680–688 (2017)
51. H. Wang, K. Huang, Y. Zhang, X. Chen, W. Jin, S. Zheng, Y. Zhang, P. Li, Recovery of lithium, nickel, and cobalt from spent lithium-ion battery powders by selective ammonia leaching and an adsorption separation system. ACS Sustainable Chem. Eng. **5**(12), 11489–11495 (2017)
52. B. Swain, Cost effective recovery of lithium from lithium ion battery by reverse osmosis and precipitation: a perspective. J. Technol. Biotechnol. **93**(2), 31–319 (2018)
53. S.H. Joo, S.M. Shin, D. Shin, C.H. Oh, J.P. Wang, Extractive separation studies of manganese from spent lithium battery leachate using mixture of PC88A and versatic 10 acid in kerosene. Hydrometallurgy **156**, 136–141 (2015)
54. Y. Pranolo, W. Zhang, C.Y. Cheng, Recovery of metals from spent lithium-ion battery leach solutions with a mixed solvent extractant system. Hydrometallurgy **102**, 37–42 (2010)
55. X. Zhang, L. Li, E. Fan, Q. Xue, Y. Bian, F. Wu, R. Chen, Toward sustainable and systematic recycling of spent rechargeable batteries. Chem. Soc. Rev. **47**(19), 7239–7302 (2018)
56. C. Lupi, M. Pasquali, Electrolytic nickel recovery from lithium-ion batteries. Miner. Eng. **16**, 537–542 (2003)
57. F. Wang, F. He, J. Zhao, N. Sui, X. Lin, H. Liu, Extraction and separation of cobalt(II), copper(II) and manganese(II) by Cyanex272, PC-88A and their mixtures. Sep. Purif. Technol. **93**, 8–14 (2012)
58. X. Chen, Y. Chen, T. Zhou, D. Liu, H. Hang, S. Fan, Hydrometallurgical recovery of metal values from sulfuric acid leaching liquor of spent lithium-ion batteries. Waste Manage. **38**, 349–356 (2015)
59. F. Wang, R. Sun, X. Jun, Z. Chen, M. Kang, Recovery of cobalt from spent lithium ion batteries using sulphuric acid leaching followed by solid-liquid separation and sovent-extraction. RSC Adv. **6**, 85303–85311 (2016)
60. A. Cerpa, F.J. Alguacil, Separation of cobalt and nickel from acidic sulfate solutions using mixtures of di(2-ethylhexyl)phosphoric acid (DP-8R) and hydroxyoxime (ACORGA M5640). J. Chem. Technol. Bietechnol. **79**, 455–460 (2004)
61. T. Suzuki, T. Nakamura, Y. Inoue, M. Niinae, J. Shibata, A hydrometallurgical process of the separation of aluminum, cobalt, copper and lithium in acidic sulfate media. Sep. Purif. Technol. **98**, 396–401 (2012)
62. S.H. Joo, D. Shin, C. Oh, J.P. Wang, S.M. Shin, Extraction of manganese by alkyl monocarboxylic acid in a mixed extractant from a leaching solution of spent lithium-ion battery ternary cathodic material. J. Power Sources **305**, 175–181 (2016)
63. E.M. Garcia, J.S. Santos, E.C. Pereira, M.B.J.G. Freitas, Electrodeposition of cobalt from spent Li-ion battery cathodes by the electrochemistry quartz crystal microbalance technique. J. Power Sources **185**, 549–553 (2008)
64. C. Lupi, D. Pilone, Electrodeposition of nichek-cobalt alloys: The effect of process parameters on energy consumption. Miner. Eng. **14**(11), 1403–1410 (2001)
65. E.M. Garcia, H.A. Tar, T. Matencio, R.Z. Domingues, J.A. dos Santos, R.V. Ferreira, E. Lorencon, D.Q. Lima, M.B. de Freitas, Electrochemical recycling of cobalt from spent cathodes of lithium-ion batteries: its application as supercapacitor. J. Appl. Electrochem. **42**, 361–366 (2012)
66. E.M.S. Barbieri, E.P.C. Lima, S.J. Cantarino, M.F.F. Lelis, M.B.J.G. Freitas, Recycling of spent ion-lithium batteries as cobalt hydroxide, and cobalt oxide films formed under a conductive glass substrate, and their electrochemical properties. J. Power Sources **269**, 158–163 (2014)
67. G.P. Nayaka, J. Manjanna, K.V. Pai, R. Vadavi, S.J. Keny, Recovery of valuable metal ions from the spent lithium-ion battery using aqueous mixture of mild organic acids as alternative to mineral acids. Hydrometallurgy **151**, 73–77 (2015)
68. S. Castillo, F. Ansart, C. Laberty-Robert, J. Portal, Advances in the recovering of spent lithium battery compounds. J. Power Sources **112**, 247–254 (2002)
69. H. Zou, E. Gartz, D. Apelian, Y. Wang, A novel method to recycle mixed cathode materials for lithium ion batteries. Green Chem. **15**(5), 1183–1191 (2013)

70. G. Cai, K. Fung, M.N. Ka, Process development for the recycle of spent lithium ion batteries by chemical precipitation. Ind. Eng. Chem. Res. **53**(47), 18245–18259 (2014)
71. A. Katsiapi, P.E. Tsakiridis, P. Oustadakis, S. Agatzini-Leonardou, Cobalt recovery from mixed Co-Mn hydroxide precipitates by ammonia-ammonium carbonate leaching. Miner. Eng. **23**, 643–651 (2010)
72. Li Li, Chen Renjie, Wu Zhang Xiaoxiao, Ge Jing Feng, Xie Man, Preparation and electrochemical properties of re-synthesized $LiCoO_2$ from spent lithium-ion batteries. Chin. Sci. Bull. **57**(32), 4188–4194 (2012)
73. L. Yao, Y. Feng, G. Xi, A new method for the synthesis of $LiNi_{1/3}Co_{1/3}Mn_{1/3}$ from waste lithium ion batteries. RSC Adv. **5**, 44107–44114 (2015)
74. Y. Lu, H. Yao, G. Xi, F. Yong, Recycling and synthesis of $LiNi_{1/3}Co_{1/3}Mn_{1/3}O_2$ from waste lithium ion batteries using D. L-malic Acid. RSC Adv. **6**, 17947–17954 (2016)
75. Q. Sa, E. Gratz, M. He, L. Wenquan, D. Apelian, Y. Wang, Synthesis of high performance LiNi1/3Mn1/3Co1/3O2 from lithium ion battery recovery stream. J. Power Sources **282**, 140–145 (2015)
76. Q. Sa, E. Gratz, J.A. Heelan, S. Ma, D. Apelian, Y. Wang, Synthesis of diverse LiNixMnyCozO2 cathode materials from lithium ion battery recovery stream. J. Sustainable Met. **2**, 248–256 (2016)
77. Y. Yang, G. Huang, M. Xie, X. Shengming, Y. He, Synthesis and performance of spherical $LiNi_xCo_yMn_{1-x-y}O_2$ regenerated from nickel and cobalt scraps. Hydrometallurgy **165**, 358–369 (2016)
78. X. Zhang, Y. Bian, X. Siwenyu, E. Fan, Q. Xue, Y. Guan, W. Feng, L. Li, R. Chen, Innovative application of acid leaching to regenerate $Li(Ni_{1/3}Co_{1/3}Mn_{1/3})O_2$ cathodes from spent lithium-ion batteries. ACS Sustainable Chem. Eng.g **6**(5), 5959–5968 (2018)
79. L. Li, X. Zhang, R. Chen, T. Zhao, L. Jun, F. Wu, Khalil Amine, Synthesis and electrochemical performance of cathode material $Li_{1.2}Co_{0.13}Ni_{0.13}Mn_{0.54}O_2$ from spent lithium-ion batteries. J. Power Sources **249**, 28–34 (2014)

Pyrometallurgical Routes for the Recycling of Spent Lithium-Ion Batteries

Huayi Yin and Pengfei Xing

1 Introduction

Renewable energy is the final solution to mitigate climate change resulted from the burning of fossil fuels that produces large quantities of greenhouse gases [1–3]. Among various forms of renewable energies, wind and solar represent the most promising candidates in the next few decades [4]. However, the wind and solar energies are intrinsically intermittent, i.e., the sun does not always shine and the wind does not always blow. To solve the intermittency, electrochemical energy storage (EES) technologies could serve as a bridge to fill the gap between the energy supply and consumption [5]. Currently, lithium-ion batteries (LIBs) are the most widely used batteries in portable devices, electric vehicles, and grid-energy storage due to their high-energy and high-power densities, simple operation, and relatively long lifetime [6–9]. As the demands for the electrified transportation and grid-energy storage power stations are increasing, the production of LIBs is continuously increasing. As a result, the fate of the end-of-life batteries is gradually attracting more attention because the spent batteries contain rare metals and toxic components that should be recycled to maintain the resource sustainability and avoid environmental pollution [10, 11].

There are three typical ways to recycle spent LIBs such as biometallurgical, hydrometallurgical, and pyrometallurgical processes [12–14]. The focus of this chapter is to overview the pyrometallurgical processes including the physical extraction process, thermodynamics of thermochemical reactions, metal solutions and slags, metal separation and refining, and furnaces [15, 16]. Pyrometallurgy primarily refers to a branch of extractive metallurgy which involves the thermal treatment of minerals, ores, and concentrates to recover valuable metals. Currently, the pyrometallurgy is dominating the extraction method of producing metals in large scales like iron and steels. Thus, many principles, experiences, and industrial equipment can be

H. Yin (✉) · P. Xing
School of Metallurgy, Northeastern University, Shenyang, People's Republic of China
e-mail: yinhy@smm.neu.edu.cn

L. An (ed.), *Recycling of Spent Lithium-Ion Batteries*,
https://doi.org/10.1007/978-3-030-31834-5_3

directly employed from the conventional pyrometallurgical fields and applied for the recycling of spent LIBs. Due to its simplicity and high productivity, the pyrometallurgical process is usually employed for industrial-scale applications [4].

The pyrometallurgical routes to recycle spent LIBs consist of two major approaches: (1) regeneration of electrode materials by lithiation or crystal repairs through a heat-treatment process, and (2) convert spent batteries into Fe-, Co-, Ni-, and Mn-based liquid alloys at a temperature higher than 1000 °C [17]. The regeneration/repair of the electrode materials is usually conducted by a heat-treatment process in which more lithium salt is supplied to recover the original constituent, and the heating process can promote the recrystallization of the electrode materials [12]. Currently, the pyrometallurgical process has been adopted to recycle spent LIBs by several companies [18–20]. Accurec GmbH (Germany) process is to recover Co–Mn alloys in a high-temperature furnace where the pretreated electrode materials in the form of briquettes are smelted to produce the Co–Mn alloy and Li-containing slag (Fig. 1) [21]. Sony/Sumitomo (Japan) process is to burn off the inflammable components whereas the residual metals are magnetically separated from the electrode materials and are thereafter recycled by a hydrometallurgical process [21]. Umicore process (Belgium and Sweden) is a hybrid process that firstly burns off the inflammable components and then melts the metallic components to generate the liquid Ni-, Co-, Cu- and Fe-alloys which are subsequently cooled down and separated by a hydrometallurgical route (Fig. 2). In addition to the transition metals, the Li ions will enter into slags that could be further recycled. Therefore, the combination of pyro- with hydro- processes could be able to take advantage of the high efficiency and productivity, and to tackle the challenges of high energy consumption and emitting hazardous gas.

In addition to the above-mentioned two pyrometallurgical processes, the pyrolysis of the mixture of cathode and carbon anode was studied by Li et al. [22]. This process uses carbon to reduce and break down the chemical bonds of $LiMO_x$ (M refers to transition metals) to generate Li_2CO_3 (or Li_2O and CO_2), MO_y/M, and CO_2 under vacuum or in an inert gas atmosphere. The Li_2CO_3 can be separated from MO_y or M by a water leaching process. MO_x, M, and C could be separated by a flotation or a magnetic separation method. The obtained MO_y with a relatively low valence value could be directly used as raw materials to resynthesize cathode materials by a solid-state reaction. Thus, the carbon-assisted pyrolysis eliminates the use of strong acid solutions of a typical hydrometallurgical process and high energy consumption and emissions of a conventional pyrometallurgical route, which is a promising process for large-scale applications.

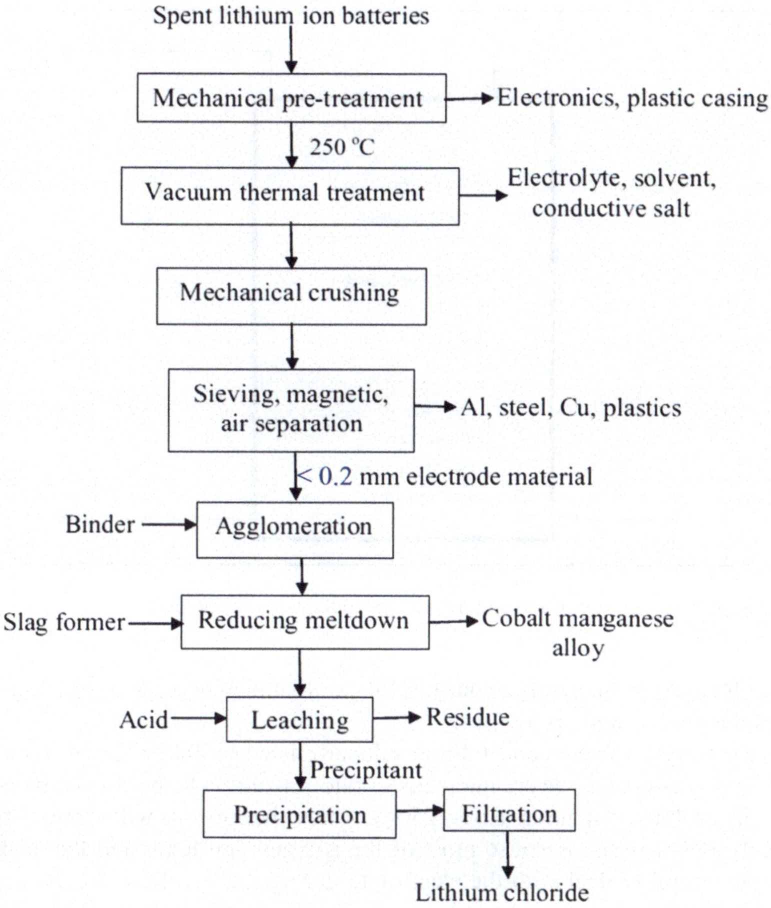

Fig. 1 Flow sheet of the Accurec recycling process [21]

2 Key Components of LIBs

To recycle spent LIBs, physiochemical properties of the components of LIBs are crucial for choosing appropriate pyrometallurgical conditions. In addition to the plastic case and outer casting, a LIB consists of a positive electrode, a separator, a negative electrode, and the electrolyte. The components of different types of LIBs are shown in Table 1 [21]. Unlike minerals, LIBs contain various materials such as plastic, metals, oxide, carbon, salts, and organic solvents. It should be pointed out that some components of LIBs are volatile, flammable, water sensitive, toxic, and unstable at high temperatures (Table 1).

The properties of plastics and metals (i.e., Cu, Al, Fe) are well known, and the separation of these materials has already been discussed by Chen et al. (Fig. 3) [4].

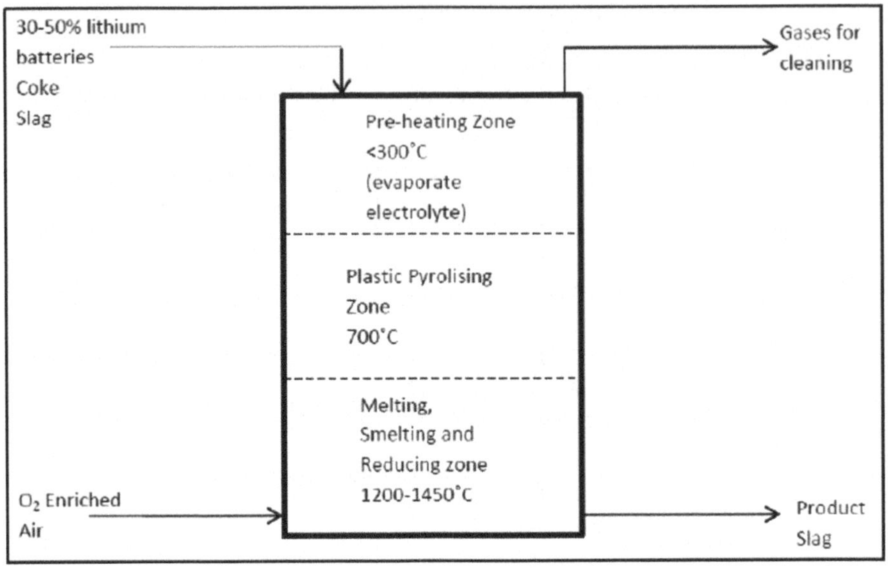

Fig. 2 Temperature zones in IsaSmelt furnace used in the Umicore process

This part focuses on the pyrolysis during both pretreatment process and the following extraction and separation processes.

The electrolyte solvents and lithium salts are listed in Tables 2 and 3. Organic carbonate and esters are the common electrolyte solvents. The boiling temperatures of most electrolyte solvents are below 250 °C. Thus, the solvents will vaporize before the pyrolysis reactions. Because most of the solvents are toxic and flammable, it should be careful to deal with the electrolyte during the pyrolysis. Many methods have been described to recycle and remove the electrolyte solvents [23–25]. The lithium salts will leave behind after the solvents are removed. As shown in Table 3 [26], most Li-salts melt and decompose above 300 °C. The decomposition of Li-salts will generate toxic fluoride compounds and LiF. Precautions should be considered to handle the toxic gas, and the LiF will enter into the slag in the following reduction smelting process.

The conductive carbon and binders should be considered in the recycling process. The conductive carbon is often mixed with the electrode active materials, and it acts as the reducing agent in the roasting or smelting process or it is burned off in the calcination process. There are various types of binders in LIBs. Polyvinylidene fluoride (PVDF) is the most commonly used binder in commercial LIBs. Note that the thermo-decomposition of PVDF will generate HF which is harmful to both human health and the environment. Rather than decomposition, dissolving PVDF into an organic solvent is another way to remove the binders [27]. In addition to PVDF, many types of binders are applied in different electrode materials. For example, the binders used in the Si-based anode could be divided into four categories: linear type, branched type, crosslinking type, and self-healing type [28]. During the pyro-treatment process,

Table 1 Various components of LIBs, electrode materials, and the properties/merits of the materials [21]

Component	wt% of the total weight of the battery	Material	Structure	Properties/merits
Cathode	39.1 ± 1.1	$LiCoO_2$	Layered	High structural stability and can be cycled for >500 times with 80–90% capacity retention
		$LiMn_2Q_4$	Spinel	Attractive for ecological and economic reasons; discharges – 3 V
		$LiNiO_2$	Layered	Cheaper and possesses higher energy density (15% higher by volume, 20% higher by weight), but less stable and less ordered as compared to $LiCoO_2$
		$LiFePO_4$	Olivine	Suitable for biomedical applications because of higher safety levels and lower cost
		Li_2FePO_4F	Olivine	
		$LiCo_{1/3}Ni_{1/3}Mn_{1/3}O$	Layered/spinel	Possesses high capacity with structural and thermal stability, and safe to use
		$Li(Li_2Ni_x Mn_yCo_z)O_2$	Layered/spinel	

(continued)

Table 1 (continued)

Component	wt% of the total weight of the battery	Material	Structure	Properties/merits
Anode		Carbon	Graphite	Low cost and availability. It has the ability to reversibly absorb and release large quantity of Li (Li:C = 1:6)
		Hard carbon	Microspheres	
Electrolyte		Lithium salt like $LiPF_6$, $Li[PF_3(C_2F_5)_3]$ or $LiBC_4O_8$ in organic solvents		Withstands high temperature and possesses high mobility of Li ions
Plastic case	22.9 ± 0.7	Polyethylene terephthalate layers, a polymer layer and a polypropylene layer, layers of carbonized plastic		Hermetically sealed battery body which converts chemical energy to electrical energy in order to generate current
Outer casing	10.5 ± 1.1	Stainless steel, aluminium		
Copper foil	89 ± 03	Copper		−14 μm thick
Aluminium, foil	6.1 ± 0.6	Aluminium		−20 μm thick
Polymer foil and electrolyte	5.2 ± 0.4	Polyethylene, polypropylene or composite polyethylene/polypropylene films		Use of 3–8 μm layers (PP/PE/PP) with 50% porosity
Solvent	4.7 ± 0.2	Ethylene carbonate, dimethyl carbonate and diethyl carbonate		Non-aqueous
Electrical contact	20 ± 05	Aluminium and copper		Conductive

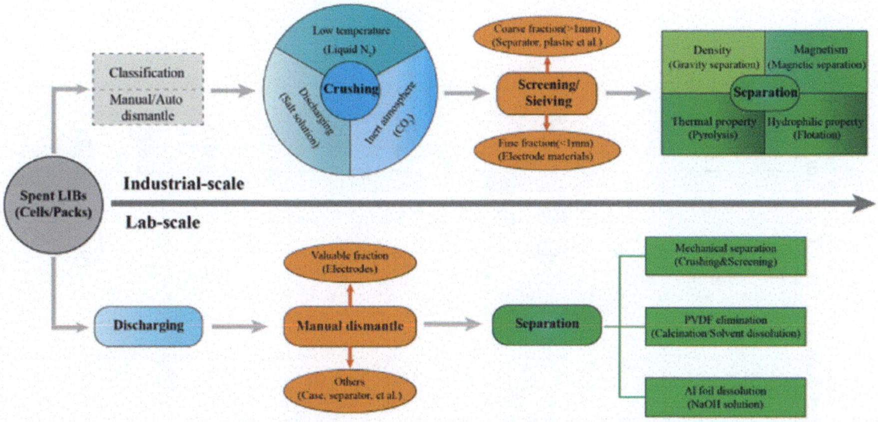

Fig. 3 Industrial-scale and lab-scale pretreatment processes and technologies of spent LIBs [4]

the organic binders are decomposed to hydrogen-containing compounds (e.g., HF, H_2O, etc.), C and CO_2. It should be aware of toxic emissions during the pyrolysis (Table 4).

3 Calcination and Roasting

3.1 Calcination

Generally, calcination is a thermal treatment process for solid materials that involve thermal decomposition, phase transition, or removal of the volatile substance in the absence or limited supply of air or oxygen. In the process of metallurgy, calcination is a common method to process minerals that can be decomposed to generate a desirable compound. For example, most oxides could be prepared by calcinating carbonate minerals in the air (Eqs. 1, and 2).

$$\text{Solid A} = \text{Solid B} + \text{Gas} \tag{1}$$

$$MCO_3(s) = MO\ (s) + CO_2(g) \tag{2}$$

where M refers to metals. For treating materials from spent LIBs, lithium salts (e.g., $LiPF_6$, Li_2CO_3), organic binders (e.g., PVDF, SBR), and transition metal oxide compounds (e.g. Li_xMO_2, Li_xMPO_4) could be decomposed to various materials at elevated temperatures in the air or under vacuum(Eqs. 3–6).

$$LiPF_6 = LiF + PF_5(g) \tag{3}$$

Table 2 Organic carbonates and esters as electrolyte solvents [26]

Solvent	Structure	M. Wt	T_m/°C	T_b/°C	η/cP 25 °C	ε 25 °C	Dipole moment/debye	T_ρ/°C	d/gem^{-3}, 25 °C
EC	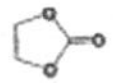	88	36.4	248	1.90, (40 °C)	89.78	4.61	160	1.321
PC	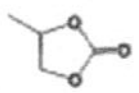	102	−48.8	242	2.53	64.92	4.81	132	1.200
BC		116	−53	240	3.2	53			
γBL	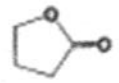	86	−43.5	204	1.73	39	4.23	97	1.199
γVL	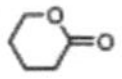	100	−31	208	2.0	34	4.29	81	1.057
NMO	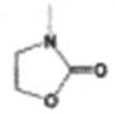	101	15	270	2.5	78	4.52	110	1.17
DMC	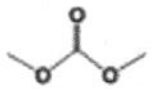	90	4.6	91	0.59 (20 °C)	3.107	0.76	18	1.063
DEC	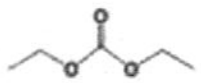	118	−74.3[a]	126	0.75	2.805	0.96	31	0.969
EMC	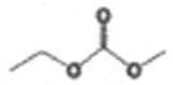	104	−53	110	0.65	2.958	0.89		1.006
EA	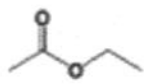	88	−84	77	0.45	6.02		−3	0.902
MB	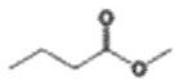	102	−84	102	0.6			11	0.898
EB	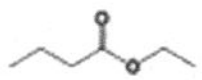	116	−93	120	0.71			19	0.878

Table 3 Lithium salts as electrolyte solutes [26]

Salt	Structure	M. Wt	T_m/°C	$T_{decompose}$/°C in solution	Al-corrosion	σ/mScm^{-1} (1.0 M, 25 °C)	
						in PC	in EC/DMC
LLBF$_4$	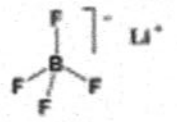	93.9	293	>100	N	3.4	4.9
LiPF$_6$	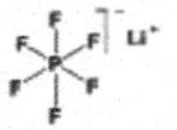	151.9	200	~80 (EC/DMC)	N	5.8	10.7
Li AsiF$_6$	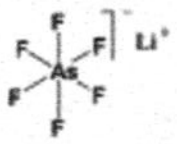	195.9	340	>100	N	5.7	11.1
LiClO$_4$	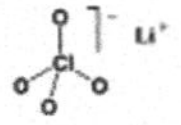	106.4	236	>100	N	5.6	8.4
Li Triflate	Li^+ $CF_3SO_3^-$	155.9	>300	>100	Y	1.7	9.0
Li Imide	Li^+ $[N(SO_2CF_3)^+$	286.9	234	>100	Y	5.1	
Li Beti	Li^+ $[N(SO_2CF_2CF_3)_2]^+$				N		

Table 4 Polymer architectures of binders for Si-based anodes [2]

	Linear-type		Branched-type	Crosslinking	Self-healing
	Homopolymer	Copolymer			
Polymer chemistry				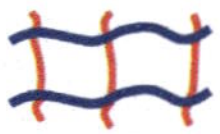	
	Identical unit	Two or more different monomer units	Formed by the replacement of a substituent	3D Crosslinked structure based on chemical covalent bond	3D Crosslinked structure based on hydrogen bond
Binders	PVDF, CMC, PVA, PA A, PAA-Na	SBR, alginate, P(AA-co-VA)	β-cyclodextrin (β-CDp)	PAA-CMC, PVA-PAA, c-PAA	
Function	No/weak interaction between a binder and Si	Weak interaction between a binder and Si	Strong interaction between a binder and Si	Covalent bond between a binder and Si	Recovering binder-Si or binder-binder interaction

$$Li_2CO_3 = Li_2O + CO_2(g) \tag{4}$$

$$PVDF = C + HF\ (g) \tag{5}$$

$$2LiMO_2 = Li_2O + M_2O_3 \tag{6}$$

The tendencies of decomposition reactions could be predicted by their thermodynamic data. It is noted that the calcination process could generate toxic gases, i.e., PF_5, HF, so that a post-treatment or an off-gas capturing process should be installed to avoid contaminations. Reactions (3–6) are only examples representing the reactions in the process of calcination, more complex reactions could exist in a practical process. The liquid electrolyte solvents are not considered in the calcination process because the liquid organic solvents are so volatile and do not exist at the temperature when the calcination reactions take place.

3.2 *Roasting*

A roasting process is carried out by heating the substance in the air or in oxygen, which is widely used in unit operation for treating the sulfide minerals [29]. Since spent LIBs contain various materials, the roasting process could burn off carbon, organic substances, and some metals. Before the roasting process, the components of the as-received materials after pre-treatment should be identified. A roasting process will remove all of the organic materials and carbons. Thus, the recycling of organic materials or carbon should be conducted before the roasting process.

A roasting process can be also used to extract cathode materials by a chemical reaction. For example, the sulfation roasting was applied to treat $LiCoO_2$ [29].

$$\begin{aligned} 4LiCoO_2(s) + 12NaHSO_4(s) &= 6Na_2SO_4(s) \\ &+ 4CoSO_4(s) + 2Li_2SO_4(s) + 6H_2O\ (g) + O_2(g) \end{aligned} \tag{7}$$

As presented in Eq. (7), the use of $NaHSO_4$ is to break down the chemical bonds of $LiCoO_2$, resulting in the production of the soluble $CoSO_4$ which could be dissolved and separated in an aqueous solution. The roasting process happened at elevated temperatures greatly improves the kinetics of the reaction. In addition to $NaHSO_4$, other inorganic salts could be used as roasting reagents, and the feasibility of the reaction depends on the Gibbs free energy change of the reactions [30, 31].

4 Oxide Reduction and Smelting

4.1 Oxide Reduction

After calcination and roasting processes, most of the residual materials are lithium metal oxides (LMOs) and some metal scraps coming from the outer casting and current collectors. If the metal scraps are separated in the pretreatment process, the residue is the LMOs. The next goal is to separate the Li ions from LMOs. It is possible to decompose LMOs to generate lithium oxide and metal oxides at high temperature, but it requires much higher energy consumption and the separation of Li_2O and metals oxides is difficult in the subsequent extracting process.

Recently, Xu et al. [32] reported a vacuum pyrolysis approach utilizing carbon to reduce $LiCoO_2$. After carbothermic reduction, $LiCoO_2$ was converted to CoO or Co and Li_2CO_3 under vacuum at high temperatures (Fig. 4). This method was also applied to treat $LiMn_2O_4$ and $LiCo_xMn_yNi_zO_2^{32}$. The use of carbon is able to reduce LMOs and thereby destroy the oxygen framework. As a result, Li_2O is liberated from the oxygen framework and then combines with CO_2 to form Li_2CO_3. After pyrolysis, Li_2CO_3 can be leached out from the solid in water, and the Co powder is collected by a magnetic separation approach [22]. Note that the components of the obtained products depend on the operating temperatures as well as the kinetics of the reaction that is governed by the operating conditions.

As shown in Fig. 5, the order of stability of various oxides could be obtained by the Ellingham diagram. In principle, the element whose Ellingham line at a lower position is more reactive than the element whose Ellingham line at a top position. As a result, the elements of the bottom line are able to reduce the oxide of the elements of the top line. Thus, the oxides whose Ellingham lines locate above the reducing agents can be reduced to their elemental state. For example, Fe-, Ni-, and Co-oxides can be reduced to elemental Fe, Ni and Co by C, CO and H_2, but MnO cannot be reduced to Mn by C, Co and H_2. On the other hand, the Gibbs free energy of the reactions could be altered by altering temperature, activities of both reductant and oxidant, vapor pressure, etc. However, the thermodynamic data of the LMO_x (i.e., $LiCoO_2$, $LiMn_2O_4$, $LiNiO_2$, $LiNi_{1/3}Co_{1/3}Mn_{1/3}O_2$ and $LiFePO_4$) are limited, and more efforts should made to measure or calculate the thermodynamic data of LMO_x. Since the open circuit potentials of LMO_x with respect to Li are higher than those of their pure oxides versus Li [33], the Gibbs free energy of formation of the lithium metal oxides (LMO_x) is higher than that of their oxides. Thus, lithium metal oxides can be reduced if their corresponding oxides can be reduced by carbon. Therefore, the possibility of the reduction of LMOx by carbon could be assessed by the thermodynamics of their corresponding oxides.

More detailed information of the reduction of $LiCoO_2$ by C under vacuum was reported by Li et al. [32]. The oxygen octahedrons collapsing model was built showing that the higher affinity of graphite to oxygen than lithium and cobalt could break down the oxygen octahedrons insides the $LiCoO_2$ [34] (Fig. 6).

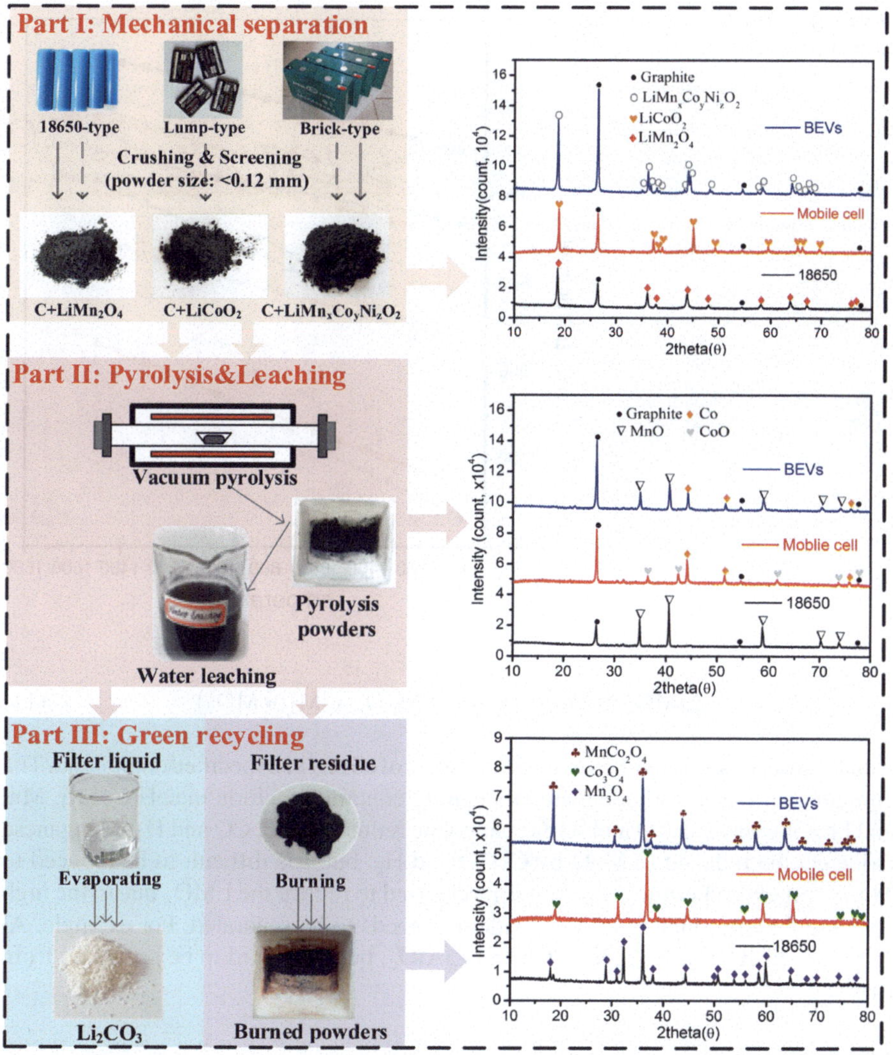

Fig. 4 The integrated process of recycling Li_2CO_3 from spent LIBs [32]

The reduction of lithium metal oxide (LMO_x) by carbon, hydrogen, CO, and reactive metals (M_R) can be expressed as follows:

$$LiMO_x + C \rightarrow Li_2CO_3 + M(\text{or } MO_y) \tag{8}$$

$$LiMO_x + CO(g) \rightarrow Li_2CO_3 + M(\text{or } MO_y) \tag{9}$$

$$LiMOx + H_2(g) \rightarrow LiOH + M(\text{or } MO_y) \tag{10}$$

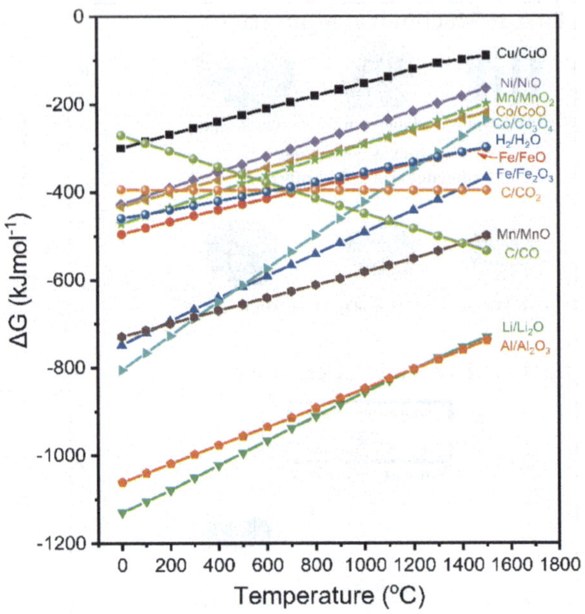

Fig. 5 The Ellingham diagram of several oxides

$$LiMO_x + M_R \rightarrow Li_2O + M_RO_z + M(\text{or } MO_y) \quad (11)$$

where subscripts x, y and z refer to the numbers of atoms in a specified compound. The commercially used cathode materials mainly contain transition metals Co, Ni, Mn, and Fe. Obviously, Li_2O and Al_2O_3 cannot be reduced by C, CO and H_2. Manganese oxide can be reduced to MnO by C, CO and H_2, but it is difficult to be reduced to Mn. The metallothermic reduction is rarely used to reduce the LMO_x due to the high cost and the introduction of new oxide that needs to be separated. For example, Al can reduce $LiCoO_2$ to generate Co and $LiAlO_2$, but Li is hard to be separated from $LiAlO_2$.

4.2 *Smelting*

Smelting is a melting process in which all constituents of the charge are in molten states and exist in two or more phases such as slags, matte, speiss, and metal. The charge usually contains fluxes, reducing agents, and minerals. The flux is often added into the charge to make a low-melting slag phase, and the reducing agent is used to convert the oxide to metals/alloys. Three types of smelting include reduction smelting, matte smelting, and flash smelting [35]. For treating spent LIBs, the commonly used approach is the reduction smelting because the feedstocks from the spent LIBs mainly consist of oxides. Maschler et al. [36] reported a hybrid pyrometallurgical and

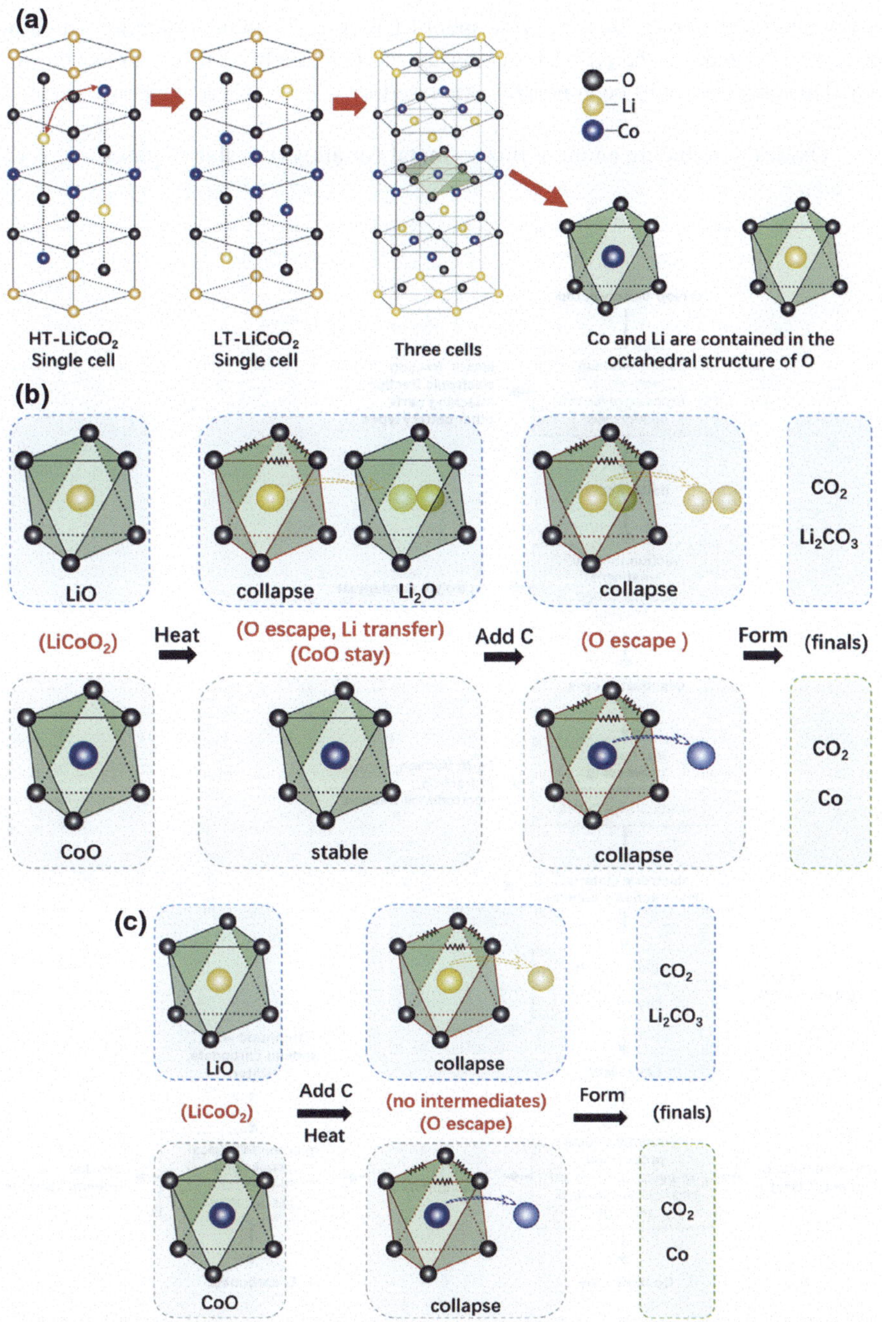

Fig. 6 Collapsing model of recycling metals from $LiCoO_2$ batteries by a roasting process, **a** Crystal structure of $LiCoO_2$ and basic cells for $LiCoO_2$, **b** Pyrolysis of $LiCoO_2$ and reduction of $LiCoO_2$ by C in order, **c** Reaction of $LiCoO_2$ and C directly [34]

hydrometallurgical process to recycle spent LIBs (Fig. 7). In the reduction smelting process, oxides are reduced by reducing agents (i.e., carbon) in the presence of a flux. This process could be generally expressed as:

$$\text{Oxides} + \text{reducing agent} + \text{flux} = \text{metal (or alloys)} + \text{slag} + \text{gases} \quad (12)$$

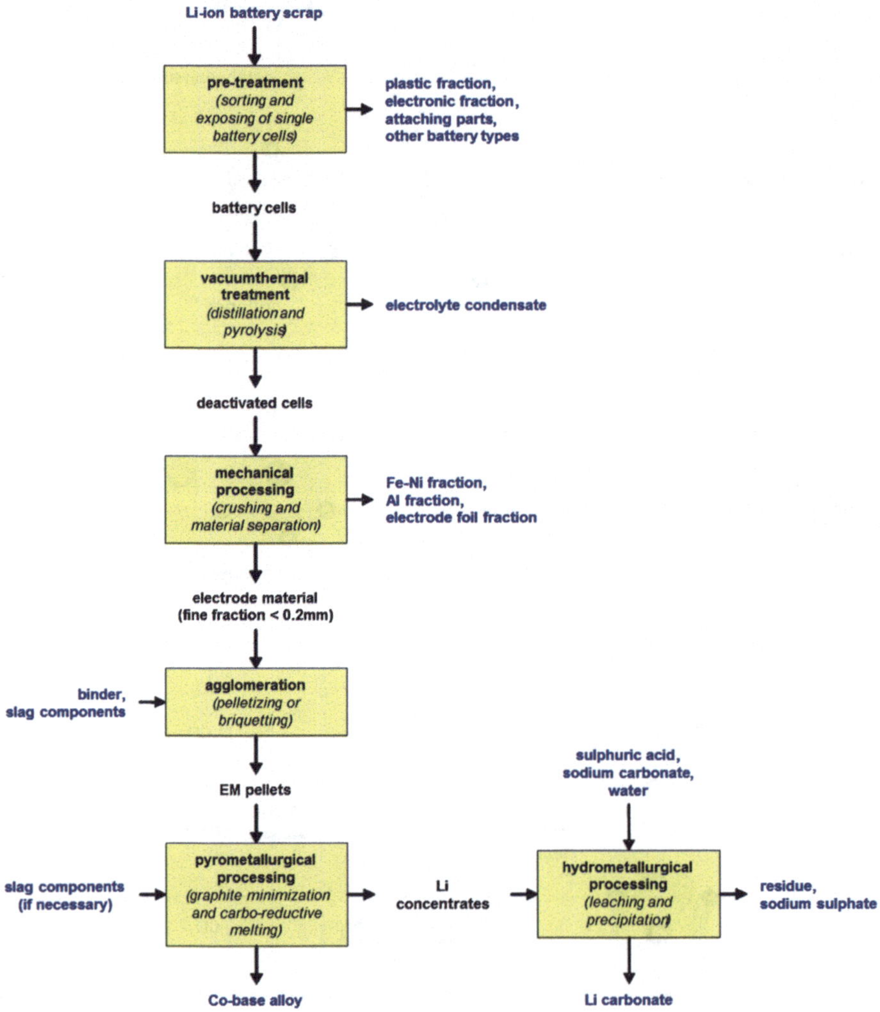

Fig. 7 A flowsheet of a combination of a pyro- hydro-metallurgical process for the recycling of portable LIBs [36]

The reduction smelting process is usually carried out in a blast furnace which will be discussed later. The oxides come from the cathode materials of the spent LIBs, i.e., $LiCoO_2$, $LiMn_2O_4$, $LiNiO_2$, $LiNi_{1/3}Co_{1/3}Mn_{1/3}O_2$, and $LiFePO_4$. The reducing agents are often carbon, CO, and natural gas. The commonly used flux contains CaO and SiO_2 which are able to adjust the fluidity and melting temperature of the slag. After the reduction completes, the transition metals enter into the liquid metal pool at the bottom of the furnace and lithium oxide stays at the top slag layer. As the slag and liquid metals are immiscible, the liquid metals are tapped from the furnace, leaving behind the slag. At last, the useful elements could be recycled by following separation procedures.

5 Metal Refining

The metals obtained from the smelting process inevitably contain impurities such as oxygen and nitrogen from the air, and carbon, aluminum and copper from the battery. The impurities not only affect the properties of the materials but also are valuable to their own right. Generally, three typical metal-refining processes are metal-slag, metal-metal, and metal-gas.

5.1 Metal-Slag Process

Due to the different reactivities of different elements, the more reactive metals in the alloys are preferentially oxidized, leaving behind the more noble metals. In other words, the less noble metal can be slagged by selective oxidation, which is also named as the fire refining. Note that the selective oxidation is applicable to treat the liquid metals having more noble metal as the bulk constituents. The as-received metals are mostly transition metals with a small amount of Al that comes from the current collector. As Al is more reactive than Fe, Co and Ni, Al could be selectively oxidized by blowing air or oxygen into the liquid metals. The oxygen affinity of different metals could be obtained by the Ellingham diagram. In some cases, the volatile oxides escape automatically and the high-melting-point oxides can be removed by skimming with the assistance of adding silica. In addition to being more reactive than the matrix metal, an impurity can be favorably removed if the impurity has a high activity coefficient in the liquid metals. Moreover, the impurity oxide can be removed if the impurity has a low activity coefficient in the slag.

5.2 Metal-Metal Process

A metal-metal process is to separate a metallic impurity from a metallic melt by phase evolution. The feasibility of the separation can be assessed by the phase diagram of the matrix metal and metal impurities. Since the temperature remarkably affects the immobility of two metals, altering the temperature of the metallic melt could result in the solidification as well as the phase separation, and thereby removing the impurities from the matrix metal. For example, the removal of copper from lead is carried out by cooling the molten copper-bearing to a temperature just above the melting point of Pb. In this scenario, the solubility of copper drastically decreases when the temperature approaches the melting point of Pb, resulting in the copper floating on the top of Pb pool and thereafter the solid Cu is separated from the liquid Pb pool. To further remove the impurities, another way is to add sulfur into the metallic melt to selectively react with impurity metal and form a molten matte. Likewise, selective alloying is able to remove impurities by adding a metal that reacts with the impurity metals and forms insoluble intermetallic. Overall, the metal-metal process is the phase separation enabled by altering the temperature or forming new phases. Although the metal-metal process is rarely applied for separating impurities from Fe-, Co-, Ni-, and Mn-alloys, this method is suitable for separating impurities from the spent liquid metals batteries (LMBs) which employ a large number of metalloids (i.e., Sb, Pb, Bi, Te, Sn) as the electrode materials [37–39]. The LMBs can be also categorized as a high-temperature LIB if the batteries use the Li^+ to carry the charge during charge and discharge.

5.3 Metal-gas Process

A metal-gas process involves the separation of impurities by distillation if the vapor pressure of the impurity is much higher than that of the matrix metal. On the other hand, the matric metal could be distilled off and collected as a pure form if the matrix metal is more volatile than the impurity metal. Sometimes, selective compounding or alloying could change the nonvolatile impurity to a volatile compound that could be separated by distillation. As we know, the use of vacuum could greatly facilitate the distillation process under a relatively low temperature. The relationship of the vapor pressure of metal with temperature is presented in Eq. 13.

$$\mathrm{Log\,P} = \mathrm{A/T} + \mathrm{B} \tag{13}$$

where P is the equilibrium vapor pressure of a metal at the temperature of T, A and B are constants. For the recycling of spent LIBs, the metal-gas procedure could be applied to remove some volatile species in the obtained metallic melt.

6 Regeneration of Cathode Materials

The electrode active materials from the spent LIBs can be regenerated by thermal treatment without dissociating the LMO through a hydrometallurgical or a pyrometallurgical process. After repeated cycles, the electrode active materials undergo the change of crystal structure and volume, resulting in the degradation in terms of specific capacity, rate capability, and efficiencies. In other words, the irreversible reactions during the cycling cause the capacity fading, meaning that partial active materials become inactive. For example, some electrode materials lose contact with the current collector, involve the phase change, or continuously react with the electrolyte. Although some of the electrode materials lose activity, the electrode materials are still inside the batteries. Rather than completely dissociating the electrode material via complex chemical leaching, precipitation and solid-state synthesis, repairing or healing the electrode material also attracts increasing attention [40]. Since a solid electrolyte interface (SEI) exists on the surface of both anode and cathode, removing the SEI is a necessary step to heal the electrode materials. Zhang et al. [41] repaired $LiCoO_2$ with a solid-state reaction between the spent electrode and Li_2CO_3, and the properties of the regenerated $LiCoO_2$ were close to the commercial $LiCoO_2$. Song et al. [42] reported a thermal treatment to repair the used $LiFePO_4$, and the sample that was treated at 650 °C showed similar performances with commercial $LiFePO_4$. The samples treated at a temperature from 600 to 800 °C showed improved tap densities and discharge capacity compared with that treated at a low temperature. Chen et al. [1] studied the different heat-treatment approaches to recycling $LiNi_{1/3}Co_{1/3}Mn_{1/3}O_2$ (NCM). The sample treated by a solvent dissolution and a following calcination process at 800 °C delivered the highest discharge capacity, and the sample treated by only calcination at 600 °C showed the best cycle performance. Chen et al. [40] proposed a combined hydrothermal with a short-annealing approach to recover $LiCoO_2$ while maintaining the original morphologies of the $LiCoO_2$ (Fig. 8).

7 Furnaces

Furnaces are indispensable for enabling various pyrometallurgical methods and providing a place for the pyro-reactions. In a lab-scale experiment, electric resistance furnace is commonly used to conduct the thermal treatments. For the industry scale application, different types of furnaces are used to process spent LIBs. As the pyrometallurgical processes for metal production are quite mature, the furnaces used for recycling spent LIBs could be adopted from the current pyrometallurgy industry.

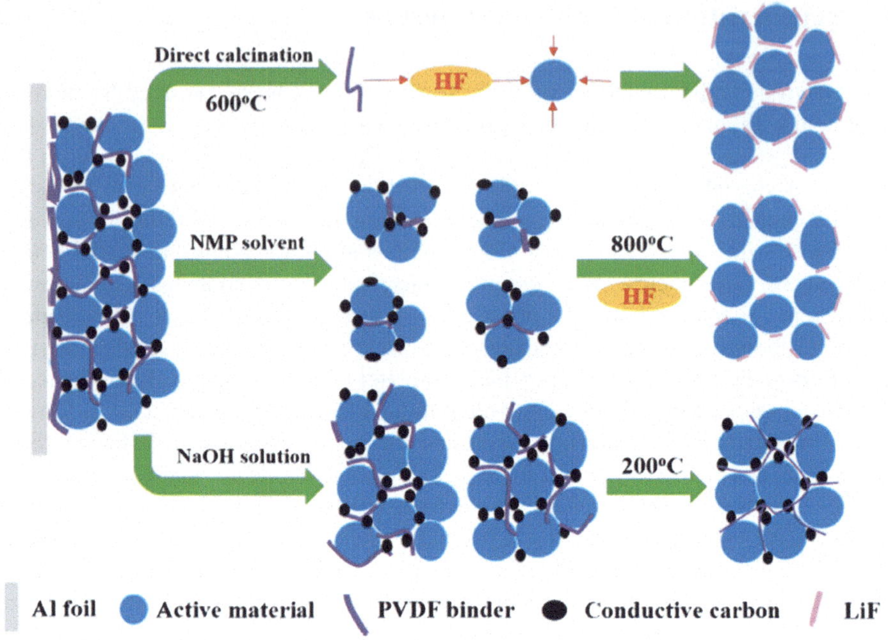

Fig. 8 Schematic illustration of different regeneration processes for $Li(Ni_{1/3}Co_{1/3}Mn_{1/3})O_2$ cathode scraps [1]

A shaft furnace is a counter-current gas-solid reactor where the flow of both gas and solid phases is counter-current. The solid charge is supplied from the upper end and gas is injected from the lower end (Fig. 9). The reactor has an upper heating zone and a lower reaction zone where the redox reactions take place. For example, the reduction of iron oxide using natural gas involves reduction, carburization, and reforming reactions. For processing battery materials, the expertise and empirical methods could be directed adopted from the reduction of iron oxides. Note that the shaft furnace is specified for treating the solid feedstocks [43].

A rotary hearth furnace has a flat, refractory hearth-rotating furnace inside a circular tunnel kiln (Fig. 10). This furnace is to enable solid-solid reactions. The pellet charge consists of oxide particles and solid reducing agents such as C and Si. The burners are used to heat the feed agglomerate and start the reduction. The fuels include natural gas, oil, or coal which are burned to generate heat that is transferred to the charge by radiation. This furnace could be used to treat the reduction of LMO by carbon [44].

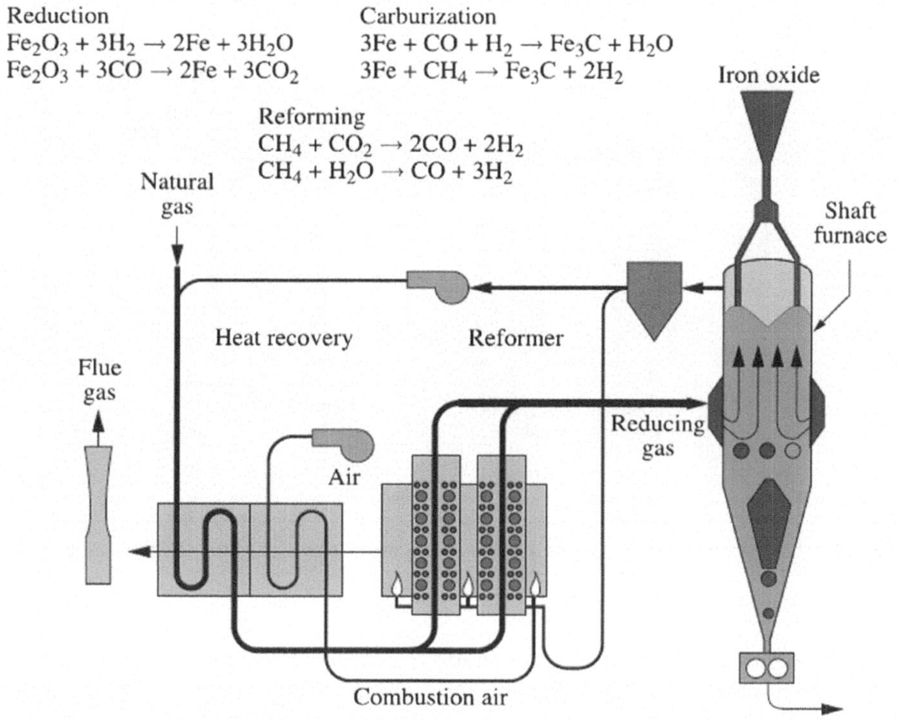

Fig. 9 Flow sheet of the Midrex process for DRI production in a shaft furnace [43]

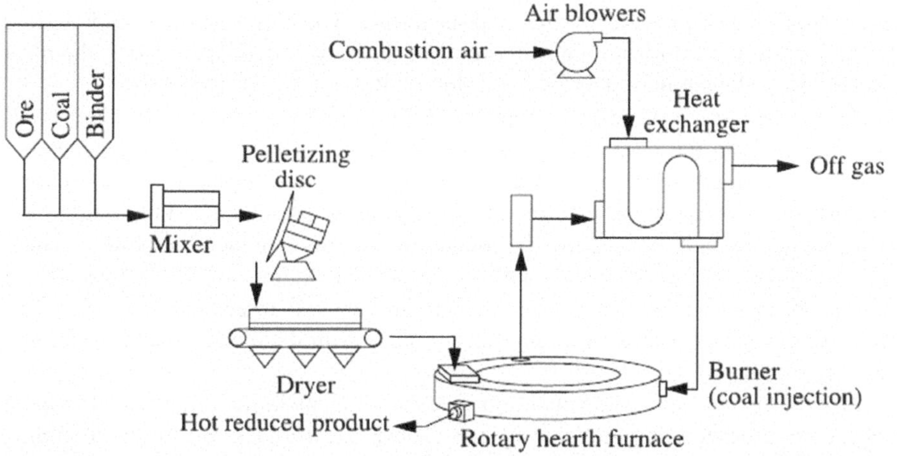

Fig. 10 Flow-sheet of the Midrex/Fasmet process for the production of DRI in a rotary hearth furnace [44]

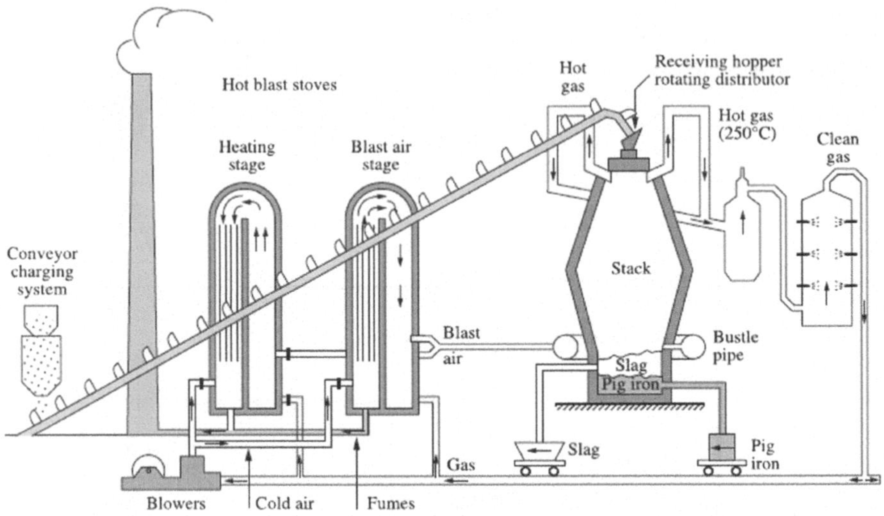

Fig. 11 The iron blast furnace and its auxiliary equipments (Arcelor-Mittal IRSID-Usinor document)

A blast furnace looks like a tall shaft-type reactor with a crucible-like hearth (Fig. 11). The solid charge is put into the furnace through the upper end, and the hot blast air is injected from the tuyeres at the bottom. A blast furnace consists of an upper counter-current gas-solid reactor for gas-solid reaction, an intermediate gas-liquid reactor (i.e., the bosh), the active zone for smelting reduction, and a lower liquid-liquid reactor for the refining and separation. The blast furnace is widely used for making liquid iron from the iron ores. For the recycling of spent LIBs, the blast furnace could be applied to prepare the liquid alloys and slags [45–48].

An electric arc furnace is to convert electric energy to thermal and mechanical energies, and these energies could be used for melting and smelting metallic charges, smelting-reduction of oxides ores. A continuous current or an alternating current passes through the electrodes, generating an arc and causing Joule effect in a conductive bath, or creating the induction currents in a metallic burden. There are various types of melting and smelting furnaces including single-electrode direct current arc furnace (Fig. 12), the three-phase furnace with three electrodes (Fig. 13), induction melting furnace, and remelting furnace with the electrode immersed in a slag layer [49].

In addition to the electrode materials in spent LIBs, metallic scraps from the spent LIBs can be treated and recycled by the electric arc furnace. From the economic viewpoint, it is cheaper to reduce the LMOs in a blast furnace than in an electric arc furnace. The electric arc furnace can be used to melt and recover the outer casting metals as well as the current collectors of the electrode materials [50].

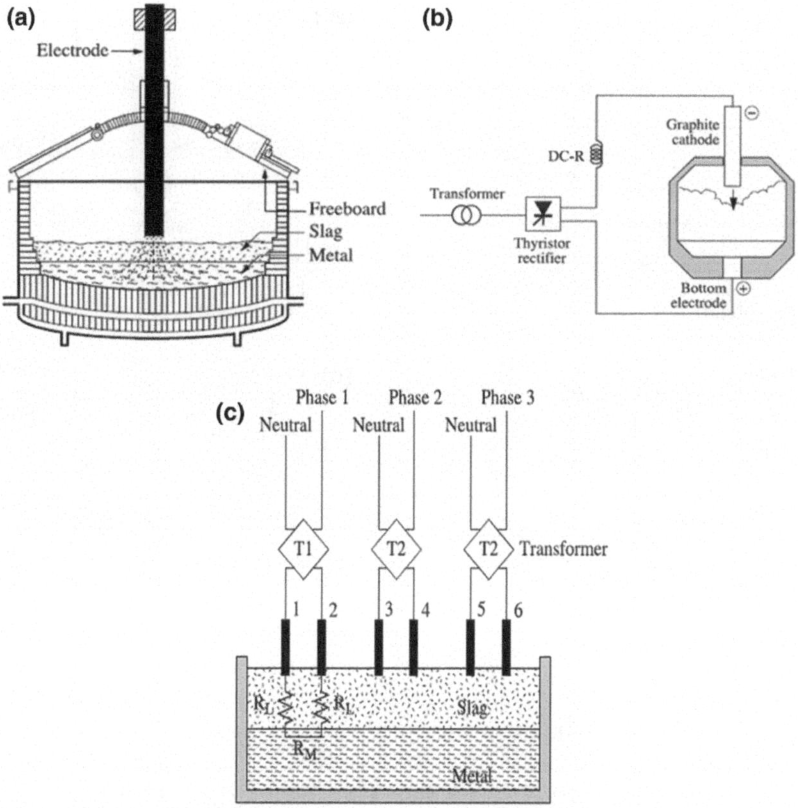

Fig. 12 **a** Single electrode direct current arc furnace, central vertical arc and bath stirring; **b** Electrical circuit of a single electrode/direct current arc furnace; **c** Electrical circuit of a six-in-line alternating current furnace with electrodes immersed in the slag layer [49]

8 Conclusions

Pyrometallurgical processes have been applied to recycle spent LIBs in industrial scale. The principles and equipment are directly adopted from the pyrometallurgical field. Although the pyro-methods are easy to scale up for industrial applications, those processes are energy intensive, generate toxic gases, and have a limited extraction rate. The pyro-process at medium temperatures (lower than 1000 °C) could be a promising method to treat the batteries materials. The combination of the pyro-methods with hydro-methods will be the central focus in the future to make a greener and more efficient process. Moreover, recycling of LIBs is an interdisciplinary subject involving chemistry, materials science, chemical engineering, metallurgy, etc. Most principles of recycling and repairing spent LIBs are adopted from the related scientific text books and current industries. However, most of the metallurgists are not familiar

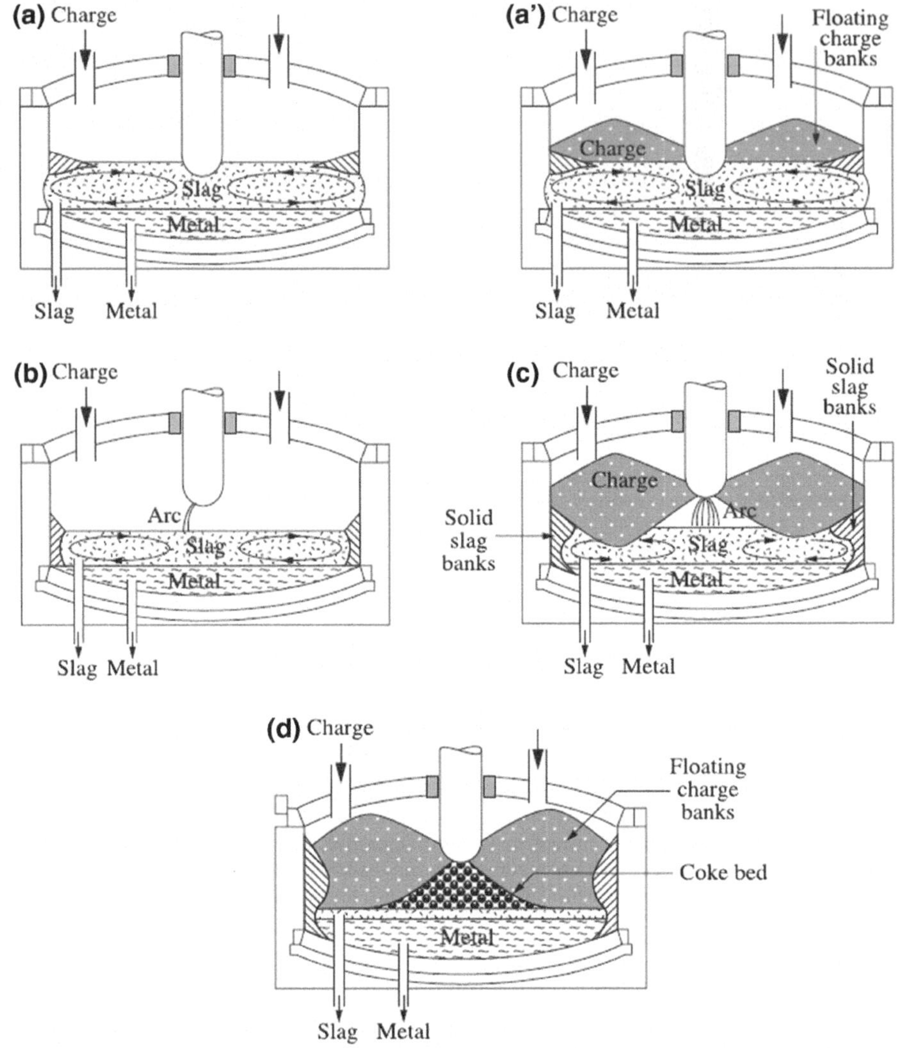

Fig. 13 Smelting reduction electric furnaces operating modes: **a** immersed electrodes in the slag and open bath; **a** immersed electrode (in the slag layer) covered by the charge; **b** open arc/open bath (short arcs); **c** shielded arc; and **d** submerged arc (in a coke bed) [50]

with the batteries. Thus, the collaboration of the scientists from battery and metallurgy fields is important to develop novel methods for recycling spent LIBs more efficiently.

References

1. X. Zhang, Q. Xue, L. Li, E. Fan, F. Wu, R. Chen, Sustainable recycling and regeneration of cathode scraps from industrial production of lithium-ion batteries. ACS Sustain. Chem. Eng. **4**, 7041–7049 (2016)
2. S. Chu, Y. Cui, N. Liu, The path towards sustainable energy. Nat. Mater. **16**, 16 (2016)
3. D. Larcher, J.M. Tarascon, Towards greener and more sustainable batteries for electrical energy storage. Nat. Chem. **7**, 19–29 (2015)
4. X. Zhang, L. Li, E. Fan, Q. Xue, Y. Bian, F. Wu, R. Chen, Toward sustainable and systematic recycling of spent rechargeable batteries. Chem. Soc. Rev. **47**, 7239–7302 (2018)
5. M.R. Palacín, G.A. De, Why do batteries fail? Science **351**, 1253292 (2016)
6. Y. Yao, M. Zhu, Z. Zhao, B. Tong, Y. Fan, Z. Hua, Hydrometallurgical processes for recycling spent lithium-ion batteries: a critical review. ACS Sustain. Chem. Eng. **6**, 13611–13627 (2018)
7. B. Huang, Z. Pan, X. Su, L. An, Recycling of lithium-ion batteries: recent advances and perspectives. J. Power Sources **399**, 274–286 (2018)
8. Y.A. Xu, Y.Y. Dong, X. Han, X.F. Wang, Y.J. Wang, L.F. Jiao, H.T. Yuan, Application for simply recovered LiCoO2 material as a high-performance candidate for supercapacitor in aqueous system. ACS Sustain. Chem. Eng. **3**, 2435–2442 (2015)
9. J.B. Dunn, L. Gaines, J. Sullivan, M.Q. Wang, Impact of recycling on cradle-to-gate energy consumption and greenhouse gas emissions of automotive lithium-ion batteries. Environ. Sci. Technol. **46**, 12704–12710 (2012)
10. X. Chen, C. Guo, H. Ma, J. Li, T. Zhou, L. Cao, D. Kang, Organic reductants based leaching: a sustainable process for the recovery of valuable metals from spent lithium ion batteries. Waste Manage. **75**, 459–468 (2018)
11. D. Pant, T. Dolker, Green and facile method for the recovery of spent lithium nickel manganese cobalt oxide (NMC) based lithium ion batteries. Waste Manage. **60**, 689–695 (2017)
12. J. Xu, H.R. Thomas, R.W. Francis, K.R. Lum, J. Wang, B. Liang, A review of processes and technologies for the recycling of lithium-ion secondary batteries. J. Power Sources **177**, 512–527 (2008)
13. X. Zhang, Y. Xie, X. Lin, H. Li, H. Cao, An overview on the processes and technologies for recycling cathodic active materials from spent lithium-ion batteries. J. Mater. Cycles Waste Manage. **15**, 420–430 (2013)
14. X. Zheng, Z. Zhu, X. Lin, Y. Zhang, Y. He, H. Cao, Z. Sun, A mini-review on metal recycling from spent lithium ion batteries. Engineering **4**, 361–370 (2018)
15. O.A. Fouad, F.I. Farghaly, M. Bahgat, A novel approach for synthesis of nanocrystalline γ-LiAlO2 from spent lithium-ion batteries. J. Anal. Appl. Pyrolysis **78**, 65–69 (2007)
16. M. Bahgat, F. Farghaly, S.A. Basir, O. Fouad, Synthesis, characterization and magnetic properties of microcrystalline lithium cobalt ferrite from spent lithium-ion batteries. J. Mater. Process. Technol. **183**, 117–121 (2007)
17. Z. Sun, H. Cao, Y. Xiao, J. Sietsma, W. Jin, H. Agterhuis, Y. Yang, Toward sustainability for recovery of critical metals from electronic waste: the hydrochemistry processes. ACS Sustain. Chem. Eng. **5**, 21–40 (2016)
18. D.C.R. Espinosa, A.M. Bernardes, J.A.S. Tenório, An overview on the current processes for the recycling of batteries. J. Power Sources **135**, 311–319 (2004)
19. L. Gaines, R. Cuenca, Costs of lithium-ion batteries for vehicles. Design (2000)
20. L.I. Jin-Hui, L.I. Xin-Hai, Y.H. Zhang, H.U. Qi-Yang, Z.X. Wang, Y.Y. Zhou, F.U. Fang-Ming, Study of spent battery material leaching process. Trans. Nonferrous Met. Soc. China **19**, 0–755 (2009)

21. P. Meshram, B.D. Pandey, T.R. Mankhand, Extraction of lithium from primary and secondary sources by pre-treatment, leaching and separation: a comprehensive review. Hydrometallurgy **150**, 192–208 (2014)
22. J. Li, G. Wang, Z. Xu, Environmentally-friendly oxygen-free roasting/wet magnetic separation technology for in situ recycling cobalt, lithium carbonate and graphite from spent LiCoO2/graphite lithium batteries. J. Hazard. Mater. **302**, 97–104 (2016)
23. D. Mu, Y. Liu, R. Li, Q. Ma, C. Dai, Transcritical CO2 extraction of electrolytes for lithium-ion batteries: optimization of the recycling process and quality-quantity variation. New J. Chem. **41**, 7177–7185 (2017)
24. Y. Liu, D. Mu, R. Li, Q. Ma, R. Zheng, C. Dai, Purification and characterization of reclaimed electrolytes from spent lithium-ion batteries. J. Phys. Chem. C **121**, 4181–4187 (2017)
25. M. Gruetzke, X. Moennighoff, F. Horsthemke, V. Kraft, M. Winter, S. Nowak, Extraction of lithium-ion battery electrolytes with liquid and supercritical carbon dioxide and additional solvents. RSC Adv. **5**, 43209–43217 (2015)
26. K. Xu, Electrolytes and interphases in Li-ion batteries and beyond. Chem. Rev. **114**, 11503–11618 (2014)
27. S.F. Lux, I.T. Lucas, E. Pollak, S. Passerini, M. Winter, R. Kostecki, The mechanism of HF formation in LiPF6 based organic carbonate electrolytes. Electrochem. Commun. **14**, 47–50 (2012)
28. N.-S. Choi, S.-Y. Ha, Y. Lee, J.Y. Jang, M.-H. Jeong, W.C. Shin, M. Ue, Recent progress on polymeric binders for silicon anodes in lithium-ion batteries. J. Electrochem. Sci. Technol. **6**, 35–49 (2015)
29. J. Lin, C. Liu, H. Cao, L. Li, R. Chen, Z. Sun, Recovery of spent lithium ion batteries based on high temperature chemical conversion. Progress Chem. **30**, 1445–1454 (2018)
30. Y. Chen, Q. Tian, B. Chen, X. Shi, T. Liao, Preparation of lithium carbonate from spodumene by a sodium carbonate autoclave process. Hydrometallurgy **109**, 43–46 (2011)
31. L.F. Medina, M.M. El-Naggar, An alternative method for the recovery of lithium from spodumene. Metall. Trans. B **15**, 725–726 (1984)
32. J. Xiao, J. Li, Z. Xu, Novel approach for in situ recovery of lithium carbonate from spent lithium ion batteries using vacuum metallurgy. Environ. Sci. Technol. **51**, 11960–11966 (2017)
33. J.B. Goodenough, Y. Kim, Challenges for rechargeable Li batteries. Chem. Mater. **22**, 587–603 (2010)
34. J. Mao, J. Li, Z. Xu, Coupling reactions and collapsing model in the roasting process of recycling metals from LiCoO2 batteries. J. Cleaner Prod. **205**, 923–929 (2018)
35. A. Vignes, *Extractive Metallurgy 1* (Wiley, London, 2011)
36. T. Georgi-Maschler, B. Friedrich, R. Weyhe, H. Heegn, M. Rutz, Development of a recycling process for Li-ion batteries. J. Power Sources **207**, 173–182 (2012)
37. H. Kim, D.A. Boysen, J.M. Newhouse, B.L. Spatocco, B. Chung, P.J. Burke, D.J. Bradwell, K. Jiang, A.A. Tomaszowska, K. Wang, W. Wei, L.A. Ortiz, S.A. Barriga, S.M. Poizeau, D.R. Sadoway, Liquid metal batteries: past, present, and future. Chem. Rev. **113**, 2075–2099 (2013)
38. K. Wang, K. Jiang, B. Chung, T. Ouchi, P.J. Burke, D.A. Boysen, D.J. Bradwell, H. Kim, U. Muecke, D.R. Sadoway, Lithium-antimony-lead liquid metal battery for grid-level energy storage. Nature. **514**, 348 (2014)
39. H. Yin, B. Chung, F. Chen, T. Ouchi, J. Zhao, N. Tanaka, D.R. Sadoway, Faradaically selective membrane for liquid metal displacement batteries. Nat. Energy **3**, 127–131 (2018)
40. Y. Shi, G. Chen, Z. Chen, Effective regeneration of LiCoO2 from spent lithium-ion batteries: a direct approach towards high-performance active particles. Green Chem. **20**, 851–862 (2018)
41. S. Chen, T. He, Y. Lu, Y. Su, J. Tian, N. Li, G. Chen, L. Bao, F. Wu, Renovation of LiCoO 2 with outstanding cycling stability by thermal treatment with Li 2 CO 3 from spent Li-ion batteries. J. Storage Mater. **8**, 262–273 (2016)
42. J. Chen, Q. Li, J. Song, D. Song, L. Zhang, X. Shi, Environmentally friendly recycling and effective repairing of cathode powders from spent LiFePO4 batteries. Green Chem. **18**, 2500–2506 (2016)
43. G.D. Hughes, Direct from Midrex 2nd Quater

44. M. McClelland, M. James, G.E. Metius, Recycling ferrous and nonferrous waste streams with FASTMET. JOM **55**, 30–34 (2003)
45. M. Burteaux, Techniques de l'Ingénieur TI-M 7400 (1992)
46. J.M.F. Himber, Chimie & industrie, génie chimique **95**, 961–975 (1966)
47. J.M.F. Himber, Chimie & industrie, génie chimique **95**, 1333–1343 (1966)
48. J.M.F. Himber, Chimie & industrie, génie chimique. **96**, 25–37 (1966)
49. Y.Y. Sheng, G.A. Irons, D.G. Tisdale, Transport phenomena in electric smelting of nickel matte: part II. Math. Model. Metall. Mat. Trans. B **29**, 85–94 (1998)
50. R.C. Francki, A.G. Matyas, K.M. Donaldson, B. Wasmund, CIM Bull. **86**, 92–99 (1993)

Bio-hydrometallurgically Treatment of Spent Lithium-Ion Batteries

Bin Huang and Jiexi Wang

Abstract With the pursuit of high energy-density and low cost lithium-ion batteries (LIBs), cobalt content in the cathode materials decreases gradually. In addition, the limited nature of lithium mineral reserves is a prominent issue as the demand for LIBs is expanding fast. Hence, traditional pyrometallurgical methods is not competent to recycle spent LIBs, because not only their economic efficiency strongly depends on the cobalt content, but also they cannot recover lithium from spent LIBs. Bio-hydrometallurgical methods for recycling spent LIBs have been proposed and attracted increasing attentions in the past decades. These methods involve biological processes in which some metabolites excreted by microorganisms are used for extraction of metals from spent LIBs. Bio-hydrometallurgy, which is a special kind of hydrometallurgy, has its unique advantages and disadvantages due to the use of living biomass. This chapter gives an overview of this method in terms of leaching mechanisms, impacts on leaching efficiency and pros and cons of it.

1 Introduction

Owing to the constant offers of upgrades in consumer electronics (CEs) and rapid development of electric vehicles (EVs), gigantic amount of lithium-ion batteries (LIBs) have been manufactured in the recent decades [1]. Moreover, the ever-increasing need for EVs will further accelerate the production of LIBs. Hence, predictably, we have to face serious disposal problems of considerable amount of spent LIBs in the near future if no appropriate recycling processes can be implemented. The impact of spent LIBs on the environment mainly involves two aspects. On one hand, most of the cathode materials contain hazardous heavy metals (e.g. Co) and

B. Huang
College of Chemistry and Bio-Engineering, Guilin University of Technology, Guilin, Guangxi 541004, China

J. Wang (✉)
School of Metallurgy and Environment, Central South University, Changsha, Hunan 410083, China
e-mail: wangjiexikeen@csu.edu.cn

L. An (ed.), *Recycling of Spent Lithium-Ion Batteries*,
https://doi.org/10.1007/978-3-030-31834-5_4

harmful organic electrolytes, which can cause serious environmental pollution [2, 3]. On the other hand, spent LIBs contain high-value metals, such as Li, Co, Ni, Cu, Al, etc. Some of them, such as Co and Li, are much more abundant in LIBs than in natural ores [4, 5]. Therefore, recycling spent LIBs is highly desirable since it can not only reduce environmental pollution, but also preserve mineral resources.

A typical process for recycling spent LIBs usually includes pretreatments and pyrometallurgical/hydrometallurgical processes which are similar to those used for recovering metals from natural ores. However, the conventional technologies have some disadvantages such as low economic efficiency, high levels of pollution and high energy consumption [6–9]. Recently, bio-hydrometallurgical methods, which are less energy intensive, eco-friendly and no emission of harmful gas, for recycling spent LIBs have been proposed and investigated intensively. Bio-hydrometallurgical methods utilize metabolites excreted by microorganisms (e.g. some bacteria and fungi) to dissolve waste electrode materials and extract valuable metals [10–12], which show similarity with traditional hydrometallurgical methods. Hence, they are regarded as a special kind of hydrometallurgical processes. To date, bio-hydrometallurgy has been a promising technology used for the recovery of metals from natural ores, industrial wastes (including spent LIBs), sewage sludge, and so on [13–15]. For example, *Aspergillus niger*, which can produce organic acids (e.g. gluconic acid, citric acid, oxalic acid and malic acid) in sucrose medium, is a promising fungus used to extract Li and Co from spent LIBs [16, 17].

2 Mechanisms of Bio-Hydrometallurgy

Bio-hydrometallurgy can be described as a process which enables dissolution of metals from their mineral sources by certain naturally inhabiting microbes [18–20]. Based on the nutrient sources of microbes, bio-leaching can be classified into two types, which are known as autotrophic (chemolithotrophic) leaching and heterotrophic leaching. Most autotrophic leachings employ chemolithotrophic, acidophilic bacteria which fix CO_2 and obtain energy from the oxidization of Fe^{2+} or reduced sulfur compounds. In the meanwhile, Fe^{3+} or H_2SO_4 can be produced in the processes. The microorganisms used in autotrophic leaching include sulfur-oxidizing bacteria (SOB), iron-oxidizing bacteria (IOB) and iron- and sulfur-oxidizing bacteria (ISOB). These bacteria can oxidize sulfur and (or) iron in leaching systems under appropriate conditions and release acidic metabolites which further solubilize the metal compounds. For example, Xin et al. [21] used mixed culture of SOB and IOB to extract Co and Li from waste $LiCoO_2$ in a $S + FeS_2$ system. The mechanism of bio-leaching can be described as Eqs. (1)–(7) (M: microbial action; C: chemical action).

$$FeS_2 + 5O_2 + 4H^+ = Fe^{3+} + 2SO_4^{2-} + 2H_2O(M) \tag{1}$$

$$FeS_2 + Fe_2(SO_4)_3 = 3FeSO_4 + 2S(C) \tag{2}$$

$$FeS_2 + 7Fe_2(SO_4)_3 + 8H_2O = 15FeSO_4 + 8H_2SO_4(C) \quad (3)$$

$$S + 1.5O_2 + H_2O = H_2SO_4(M) \quad (4)$$

$$2Fe^{2+} + 0.5O_2 + 2H^+ = 2Fe^{3+} + H_2O(M) \quad (5)$$

$$S + 3Fe_2(SO_4)_3 + 4H_2O = 4H_2SO_4 + 6FeSO_4(C) \quad (6)$$

$$2FeSO_4 + 2LiCoO_2 + 4H_2SO_4 = Fe_2(SO_4)_3 + 2CoSO_4 + Li_2SO_4 + 4H_2O(C) \quad (7)$$

In this system, the SOB oxidizes S and yields H_2SO_4 as metabolite, which can dissolute Li^+ and a part of Co^{2+} from the waste $LiCoO_2$ (Eq. (4)). Meanwhile, H^+ also promotes the production of Fe^{3+} (Eq. (1)). Subsequently, Fe^{3+} and FeS_2 can react to form Fe^{2+}, which acts as the reductant to transform insoluble Co^{3+} to soluble Co^{2+}. Under this mechanism, $LiCoO_2$ can be dissolved gradually into the leaching liquor.

For the heterotrophic leaching, however, organic carbon sources are needed by the microorganisms, whereas the metal compounds do not take part in the biological processes. In the heterotrophic process, the microorganisms usually produce mild organic acids, or non-acidic complexing agents which can be used in alkaline leaching [11]. The microorganisms used in heterotrophic are mostly filamentous fungi and bacteria. The produced organic acids can react with the metal-containing ores or industrial wastes in the leaching solution through two mechanisms which occur simultaneously or separately. Firstly, the acidic metabolites can react with the metal compounds and extract the metal cations. Secondly, the organic acids can combine with the cations to form metal-ligand complexes by chelation [16, 22].

In addition, the bioleaching processes for extraction of metals from sulfides can be classified into contact (direct) and non-contact (indirect) leaching [23], as shown in Fig. 1. For the former, microorganisms directly contacts with the metal sulfide (Fig. 1b). The microorganism development and the metal dissolution occur simultaneously. For the later (Fig. 1a), the process involves the acids generation by microorganisms and chemical leaching stage. At the beginning of the leaching stage, the microorganisms usually stop growing [11].

3 Impacts on Recovery Efficiency

Compared with traditional hydrometallurgy, bio-hydrometallurgy is simpler and more economical, but it is more difficult to be controlled because of the use of living biomass [24]. In general, the physical, chemical, and biological parameters in the bio-leaching system have significant impacts on the leaching efficiency. These

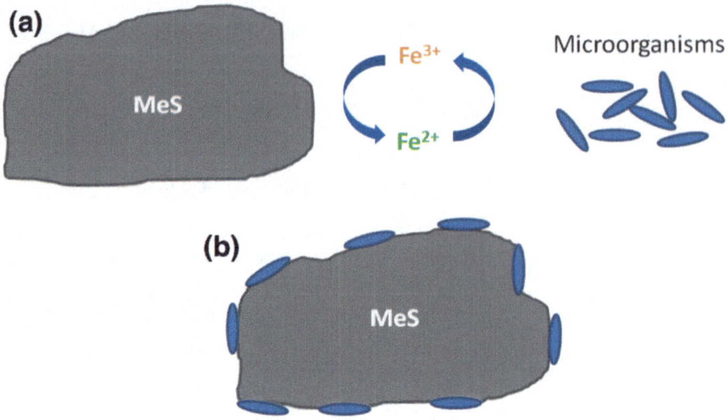

Fig. 1 **a** Non-contact (indirect) and **b** contact (direct) mechanisms

include the temperature, the pH value of the culture medium, the concentration and type of carbon source, the pulp density, the presence of trace elements, aeration, the pre-culture period and inoculum employed and the morphology of the producing microorganism [25, 26]. For example, Xin et al. [27] employed *Alicyclobacillus* sp. and *Sulfobacillus* sp. in the form of mixed culture to recover Co and Li from spent LIBs and compared the bio-leaching efficiency at different pulp densities. Their results showed that the leaching efficiency decreased from 52 to 10% for Co and from 80 to 37% for Li as the pulp density rose from 1 to 4%, which are shown in Fig. 2.

In Xin et al.'s work, the *Alicyclobacillus* sp. was used as SOB which generated H_2SO_4, while the *Sulfobacillus* sp. was used as IOB. At high pulp density, more amount of toxic organic electrolyte containing $LiPF_6$, $LiClO_4$ and $LiBF_4$ would be brought into the bioleaching media, consequently endangering the growth and activity of the mixed culture. Therefore, the SOB generated fewer amounts of H_2SO_4; in the meanwhile the IOB produced fewer amount of Fe^{3+}. As a result, the pH value of the media went higher and the oxidation-reduction potential (ORP) value decreased. Li release mainly depended on the H_2SO_4 concentration; whereas Co dissolution relied on both the SOB and IOB since the insoluble Co^{3+} in the waste cathode materials has to be reduced by Fe^{2+} into soluble Co^{2+} prior to acid dissolution [21]. Hence, the bio-leaching efficiency for both Co and Li was significantly decreased with the increase of pulp density.

In addition, Mousavi et al. [16] obtained similar results when they investigated the effect of pulp density on the bio-leaching efficiency using *Aspergillus niger*, a haploid filamentous fungus found in mesophilic environment such as decaying vegetation and soil [16]. In their work, the bio-leaching efficiency also decreased with the increase of pulp density, as shown in Fig. 3. However, the main reason for the low leaching efficiency was considered as the insufficient protons to react with the metal compounds. In Deng et al.'s work [22], it was demonstrated that higher concentration of heavy metals suppressed the activity of glucose oxidase,

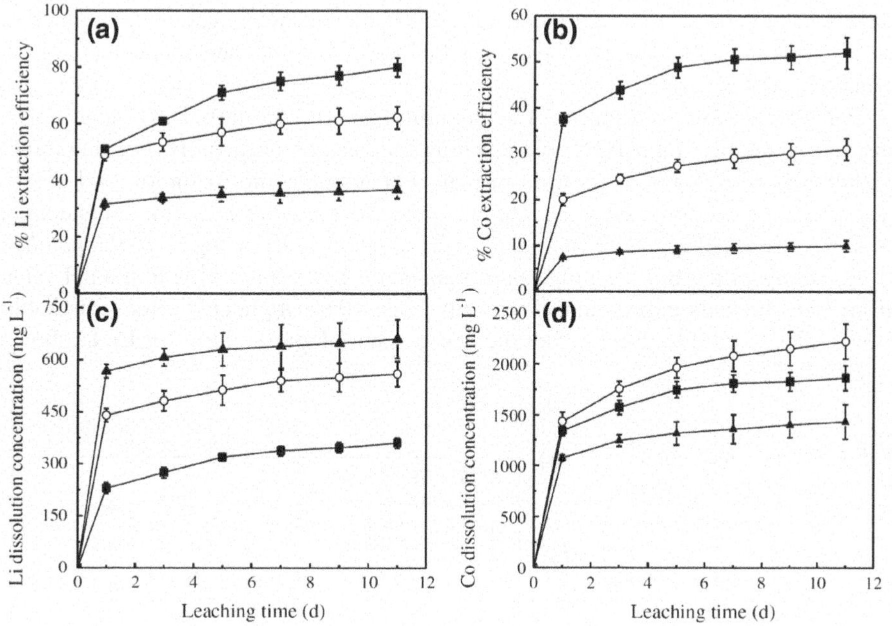

Fig. 2 Time-courses for **a** Li extraction efficiency, **b** Co extraction efficiency, **c** Li dissolution concentration and **d** Co dissolution concentration during bioleaching of spent LIBs under different pulp densities (squares 1%, circles 2%, and triangles 4%)

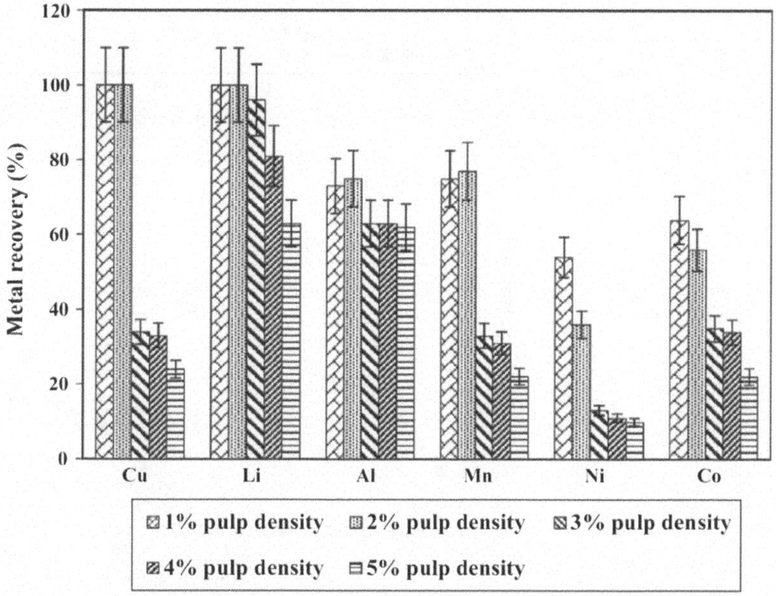

Fig. 3 metal recovery under different pulp densities in the *Aspergillus niger* bioleaching process

which was produced by fungi and could enable them to use organic compounds as nutrient sources. This result can also explain the effect of pulp density on the leaching efficiency of fungi.

For the chemolithotrophic and acidophilic bacteria, the initial pH value in the medium generally impacts the bio-leaching efficiency significantly. For one thing, higher pH value controls bacteria growth. For another, the oxidation of Fe^{2+} by bacteria will be suppressed, even inhibited at higher pH value [12, 28]. For example, Mishra et al. [12] recovered Co and Li from spent LIBs using *Acidithiobacillus ferrooxidans* as the bio-leaching microorganism. They adjusted the initial pH value from 2.5 to 4.0 in the experiment. The result showed that a higher pH value remarkably slowed the metal dissolution from the waste material, particularly for Li, as shown in Fig. 4.

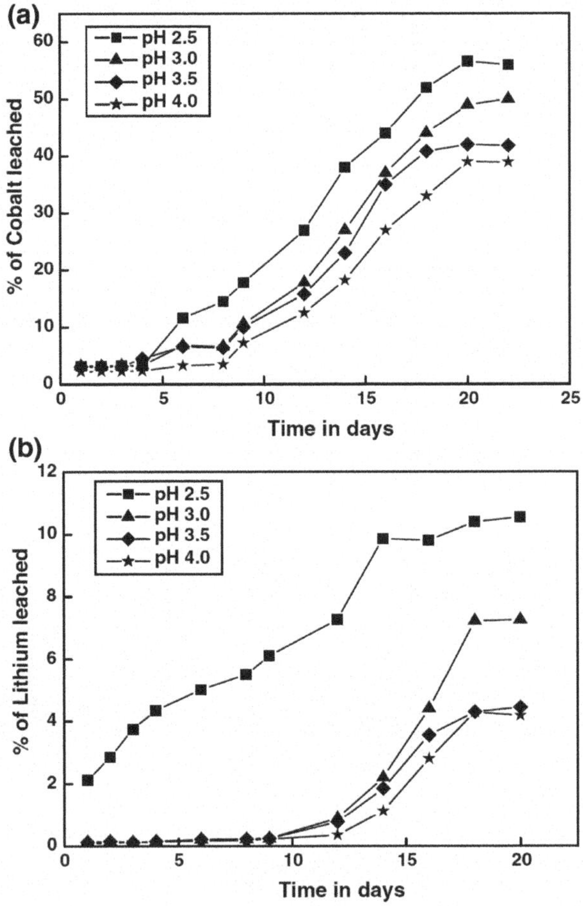

Fig. 4 Cobalt **a** and lithium **b** leaching at different initial pH values

4 Pros and Cons of Bio-Hydrometallurgy

In the past decade hydrometallurgical and bio-hydrometallurgical methods attract more attentions than pyrometallurgical one due to their low energy-consumption and less environmental pollution [7, 29, 30]. Firstly, unlike traditional pyrometallurgical method, bio-hydrometallurgical processes do not release hazardous gases (e.g. dioxins, furans etc.). Secondly, bio-hydrometallurgical processes are capable of extracting all the metals from waste electrode materials, whereas pyrometallurgical methods can hardly recover lithium and aluminum. Thirdly, compared with traditional hydrometallurgy, bio-hydrometallurgical methods do not use a lot of inorganic acids, which possess high safety risk and strict operational conditions [31]. Fourthly, bio-hydrometallurgy has lower energy consumption and lower environmental impact than traditional hydrometallurgy. In addition, bio-metallurgy is more suitable for low grade ores, mine tailings and contaminated soils.

However, because of the use of living microorganisms, bio-hydrometallurgy has some unique disadvantages. Firstly, bio-leaching processes usually require rigorous conditions since the microorganisms are sensitive to the environment changes of culture media [32]. Secondly, like traditional hydrometallurgy, bio-hydrometallurgy needs tedious recovery procedures including microbial cultivation, bio-leaching, solvent extraction, separation and precipitation [7]. Thirdly, some microorganisms produce organic acids as mild and environmental friendly leaching agents; however, organic acids may lead to difficulties in separation and recovery of metals from the leaching solution [33].

5 Summary and Outlook

In this chapter, the mechanisms, impacts on recovery efficiency and pros and cons of bio-hydrometallurgical methods for recycling spent LIBs have been summarized. Owing to the absence of high-temperature process and caustic inorganic acids, bio-hydrometallurgy is mild, environmentally-friendly and low energy-consumption. However, these methods require rigorous conditions and long operating time horizon since the cultivations of microorganisms are necessary. Based on the mechanism of bio-leaching, microbial cultivation and selection might be the most essential factor in the development of this technology.

References

1. B. Huang, Z. Pan, X. Su, L. An, J. Power Sour. **399**, 274–286 (2018)
2. H. Zou, E. Gratz, D. Apelian, Y. Wang, Green Chem. **15**, 1183–1191 (2013)
3. X. Li, J. Zhang, D. Song, J. Song, L. Zhang, J. Power Sour. **345**, 78–84 (2017)
4. T. Zhang, Y. He, F. Wang, H. Li, C. Duan, C. Wu, Sep. Purif. Technol. **138**, 21–27 (2014)

5. J. Xu, H.R. Thomas, R.W. Francis, K.R. Lum, J. Wang, B. Liang, J. Power Sour. **177**, 512–527 (2008)
6. N. Pradhan, B. Das, C.S. Gahan, R.N. Kar, L.B. Sukla, Bioresour. Technol. **97**, 1876–1879 (2006)
7. X. Chen, Y. Chen, T. Zhou, D. Liu, H. Hu, S. Fan, Waste Manage. **38**, 349–356 (2015)
8. J. Chen, R.W. Proctor, Am. J. Psychol. **128**, 197–208 (2015)
9. N. Menad, B. Björkman, E.G. Allain, Resources. Conserv. Recycl. **24**, 65–85 (1998)
10. D. Pant, D. Joshi, M.K. Upreti, R.K. Kotnala, Waste Manage. **32**, 979–990 (2012)
11. F. Anjum, M. Shahid, A. Akcil, Hydrometallurgy **117–118**, 1–12 (2012)
12. D. Mishra, D.-J. Kim, D.E. Ralph, J.-G. Ahn, Y.-H. Rhee, Waste Manage. **28**, 333–338 (2008)
13. A. Akcil, Miner. Eng. **17**, 477–480 (2004)
14. C. García-Balboa, I.C. Bedoya, F. González, M.L. Blázquez, J.A. Muñoz, A. Ballester, Bioresour. Technol. **101**, 7864–7871 (2010)
15. D.E. Rawlings, Pure Appl. Chem. **847** (2004)
16. N. Bahaloo-Horeh, S.M. Mousavi, Waste Manag **60**, 666–679 (2017)
17. W.-X. Ren, P.-J. Li, Y. Geng, X.-J. Li, J. Hazard. Mater. **167**, 164–169 (2009)
18. S. Ndlovu, Hydrometallurgy **91**, 20–27 (2008)
19. G.J. Olson, J.A. Brierley, C.L. Brierley, Appl. Microbiol. Biotechnol. **63**, 249–257 (2003)
20. D.E. Rawlings, D. Dew, C. du Plessis, Trends Biotechnol. **21**, 38–44 (2003)
21. B. Xin, D. Zhang, X. Zhang, Y. Xia, F. Wu, S. Chen, L. Li, Bioresour. Technol. **100**, 6163–6169 (2009)
22. X. Deng, L. Chai, Z. Yang, C. Tang, Y. Wang, Y. Shi, J. Hazard. Mater. **248–249**, 107–114 (2013)
23. G.M. Gadd, Geoderma **122**, 109–119 (2004)
24. X. Chen, H. Ma, C. Luo, T. Zhou, J. Hazard. Mater. **326**, 77–86 (2017)
25. T.-J. Xu, Y.-P. Ting, Enzyme Microb. Technol. **35**, 444–454 (2004)
26. B. Max, J.M. Salgado, N. Rodríguez, S. Cortés, A. Converti, J.M. Domínguez, Braz. J. Microbiol. **41**, 862–875 (2010)
27. Z. Niu, Y. Zou, B. Xin, S. Chen, C. Liu, Y. Li, Chemosphere **109**, 92–98 (2014)
28. B. Pesic, D.J. Oliver, P. Wichlacz, Biotechnol. Bioeng. **33**, 428–439 (1989)
29. J. Wang, J. Bai, J. Xu, B. Liang, J. Hazard. Mat. **172**, 1100–1105 (2009)
30. X. Chen, B. Xu, T. Zhou, D. Liu, H. Hu, S. Fan, Sep. Purif. Technol. **144**, 197–205 (2015)
31. S. Babel, D. del Mundo Dacera, Waste Manage. **26**, 988–1004 (2006)
32. Y. Liu, J. Shen, L. Huang, D. Wu, J. Hazard. Mat. **262**, 1–8 (2013)
33. A. Chagnes, B. Pospiech, J. Chem. Technol. Biotechnol. **88**, 1191–1199 (2013)

Hydrometallurgical Processes for Valuable Metals Recycling from Spent Lithium-Ion Batteries

Xiangping Chen, Ling Cao, Duozhi Kang, Jiazhu Li, Shuzhen Li and Xin Wu

Abstract A huge amount of exhausted lithium ion batteries (LIBs) will be inevitably generated with their wide application in electronic/electrical devices (e.g. mobile phones, laptops, digital cameras) and EVs/HEVs (i.e. electric vehicles/hybrid electric vehicles). And the recycling of valuable metals these spent LIBs may not only alleviate or prevent adverse impact on environment, but also expect potentially economic and resource-oriented benefits. This chapter will focus on the presentation of the relevant introduction and analysis of current hydrometallurgical processes for the separation and recovery of metals from spent LIBs. Firstly, a relatively systematic framework concerning the leaching of metals from spent LIBs is established for the discussion and evaluation of the current leaching methods, including chemical leaching and bioleaching. Then, the separation and recovery of metals from leaching solutions will be comprehensively summarized and discussed concerning chemical precipitation, solvent extraction and other separation methods. Finally, the regeneration of cathode materials by hydrometallurgical method is introduced and analyzed in this chapter. This chapter expects to present comprehensive introduction of current hydrometallurgical processes for different metals recycling from spent LIBs in aspects of leaching, metal separation/recovery and cathode materials re-preparation.

Keywords Spent lithium ion batteries · Valuable metals · Recycling · Leaching · Separation · Cathode materials regeneration

1 Introduction

In recent decades, lithium-ion batteries (LIBs) have been witnessing an increasing demand in electronic/mobile device and EV/HV due to their excellent electrochemical performances in terms of high-energy density, high voltage, long cycle life etc. [1, 2]. However, a large quantity of waste LIBs contained hazardous substances, such as heavy metals, organic pollutants, will be inevitably generated and flow into waste

X. Chen (✉) · L. Cao · D. Kang · J. Li · S. Li · X. Wu
School of Environmental Science and Engineering, Shaanxi University of Science and Technology, Xi'an 710021,, People's Republic of China
e-mail: chenxiangping101@163.com

L. An (ed.), *Recycling of Spent Lithium-Ion Batteries*,
https://doi.org/10.1007/978-3-030-31834-5_5

stream, which may cause potential risks towards the environment and human-beings without proper treatment [3, 4]. On the other hand, the sustainable recovery of valuable material, especially high value-added metals, from spent LIBs can alleviate the depletion of valuable metals and eliminate/alleviate the potentially ecological risks caused by these wastes [5, 6]. Therefore, sustainable recovery of metals from spent LIBs may expect critical significance in terms of environment pollution controlling and metal resources recycling [1, 7].

Recently, the recycling of valuable metals (e.g. Co, Li) from spent LIBs has been drawing extensive attentions [8, 9]. And the main recycling techniques can be divided into mechanical [10, 11], pyro-metallurgical [12], hydro-metallurgical [13, 14] and coupling methods [15, 16]. Among above of them, mechanical processing is a relatively simple method for the effective recovery of metals from spent LIBs with less negative environmental impacts. According to Guan et al. [17], 77% Li, 91% Co, 100% Mn and Ni can be recovered from spent LIBs by mechanochemical reduction process. However, mechanical method may be frustrated by the inapplicability of more complicated waste stream and it is also difficult to obtain relatively pure products from spent LIBs. Pyro-metallurgical process is always involved with thermal treatment. As reported by Sun and Qiu [18], vacuum pyrolysis method was adopted for the peeling off Al foils from the waste cathode materials, and high yield over 99% can be achieved under an temperature of 600 °C, processing time of 30 min and residual gas pressure of 1.0 kPa. Despite its high productivity and recovery efficiency, high-temperature involved process may increase equipment investment and aggravate secondary pollution during metals recycling. Hydro-metallurgical route (e.g. leaching/bioleaching [19, 20], solvent extraction [13], chemical precipitating [21]) has been attracting an increasing attention for its evident superiorities in aspects of higher recovering efficiency, reduced emission of hazardous substances, higher product purity over other recycling methods. Reductive leaching, selective precipitation and solvent extraction methods were adopted in our previous studies to separate and recover different metals from spent LIBs and relatively pure products can be obtained [22, 23]. Combined process (e.g. integration of mechanical pretreatment and hydrometallurgical processing) can be efficient candidate for valuable metal recycling from spent LIBs. As reported by Xiao et al. [24], an integrated process combined mechanical separation, oxygen-free roasting and water leaching was proposed to recover Li_2CO_3 (with a yield of 91.30%) from spent lithium manganese batteries ($LiMn_2O_4$). However, the hydrometallurgical and combined processes may be frustrated by tedious processes during the recycling of metals from spent LIBs. Besides, it can be also discovered that hydrometallurgical process is a welcomed choice for both the separation and recycling of relatively pure products and the consideration of environmental protection [25].

Therefore, this chapter will focus on the presentation and evaluation of the current techniques involved with hydrometallurgical processes, including leaching process, separation and recycling technologies and re-preparation of cathode materials from spent LIBs. It is acknowledged that leaching is one of the most critical steps during hydro-metallurgical recovering processes for the dissolving and extraction different metals from spent LIBs [25–27]. Besides, it can be concluded that both mineral acids (e.g. H_2SO_4 [28], HCl [29]) and organic acids (e.g. oxalic acid [30], citric acid [31]) can be used as leaching reagents. It can be also discovered that both organic acids and mineral acids illustrate similar leaching ability under the optimized conditions [25]. In this chapter, a comprehensive review of current leaching routes is firstly presented and analyzed to achieve a well-round understanding of chemical leaching and bioleaching process. After leaching, different metals should be separated or recovered respectively to obtain different products from the leaching solutions. Based on the references, current separation and recovery techniques can be divided into chemical precipitation, solvent extraction and other separation methods, such as adsorption, ion exchange and membrane separation [32–34]. The chemical precipitation process is a method using chemical reagents as precipitant for the selective precipitation or recovery of metals from leaching solutions, and relatively pure compounds can be directly recovered after precipitating reactions [28]. The solvent extraction method is a process using different organics as extractants for the extraction of metals from leaching solutions, and the loaded organics will be then stripped with acid for the subsequent purification or electrodeposition procedure [13]. Other hydrometallurgical separation and recovery techniques include adsorption, ion exchange and membrane separation [35, 36]. The adsorption method usually employs different adsorbents to adsorb and recover different metals from the leaching solutions, the ion exchange method usually uses ion-exchange resins to remove heavy metals from waste water or selectively recover metal values from leaching liquor, and membranes with specific functions are usually used to selectively separate different components from mixed feeds (e.g. liquid, gas) [13, 28, 37–41].

Therefore, a comprehensive review and analysis of the current hydrometallurgical processes is discussed in this chapter to present a clear illustration of the separation and recovery of different valuable metals from spent LIBs. Detailed hydrometallurgical processes, including the leaching process, the separation and recovery process and the cathode materials re-preparation process, are fully reviewed and discussed to reach a well-round understanding for the extraction of valuable metals from spent LIBs.

2 Leaching Processes

Usually, a typical spent LIB is composed of cathode, anode, separator, metallic shell, electrolyte and other appurtenances [35]. And the cathode and anode, made of Al and Cu foils coated with active materials (i.e. cathode materials and anode materials in the cathode and anode, respectively), are the most valuable parts during the spent LIBs recycling processes [42]. A pretreatment step (e.g. discharging, dismantling) are usually conducted in advance to remove nonmetallic ingredient and less valuable part from high value-added part before the recovery of valuable metals from spent LIBs [43]. As illustrated in Fig. 1, main pretreatment steps include discharging, removing of plastic shell, releasing electrolyte and mechanical peeling off cathode and anode. Then the cathode and anode will be subjected for the peeling off procedure

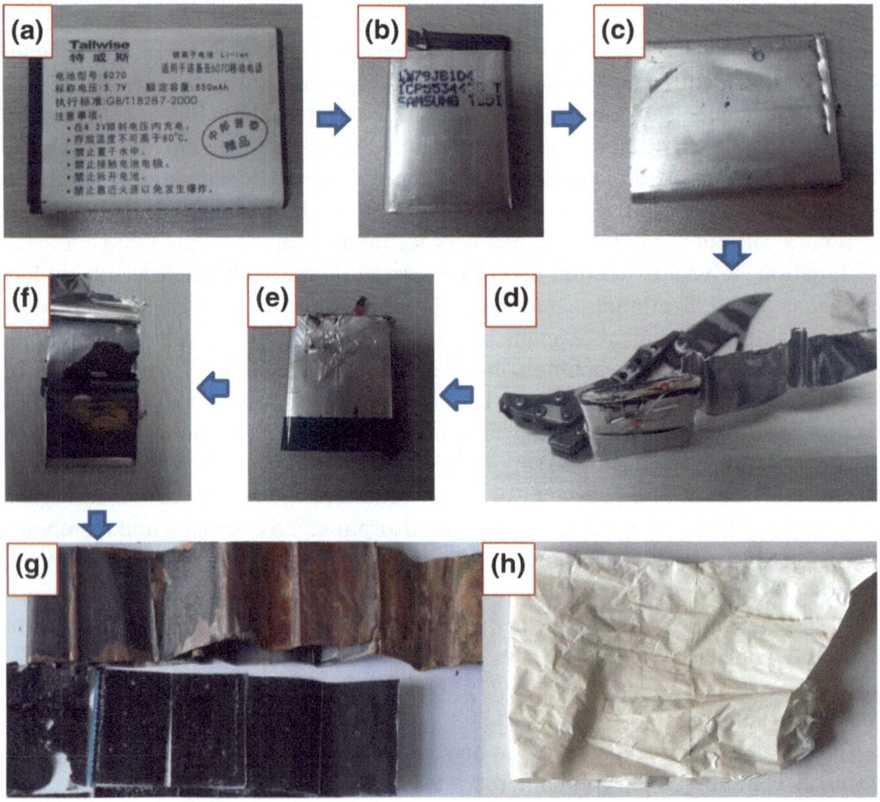

Fig. 1 Manual dismantling steps of spent LIBs. **a** The spent LIBs after discharged; **b** remove the plastic shell; **c** release electrolyte; **d**, **e**, **f** mechanical peeled off cathode and anode; **g** positive and negative current collector; **h** the film between the positive and negative components

to remove the attached Al and Cu foils, and fine particles of waste cathode materials contained high value-added metals (e.g. Co, Li) can be obtained after calcining (removing graphite and organics) and grinding (reducing particle size). Finally, the obtained particles will be used as raw materials for the subsequent leaching process. The leaching process can be categorized to two main leaching processes as chemical leaching and bioleaching, and this chapter will mainly focus on the introduction and discussion of the above two leaching processes.

2.1 Chemical Leaching

As a typical leaching method, the chemical leaching method can be an effective candidate for the dissolution of waste cathode materials from spent LIBs. Usually, different chemical reagents, including leaching reagents and reductants, are used during leaching reactions [44]. Usually, the leaching agents include mineral acids [43, 45–47] and organic acids [26, 48–50], and the reductants include inorganic reagents (e.g. hydrogen peroxide [42, 51], sodium persulfate [52]) and organic reagents (e.g. glucose [53, 54], sucrose [54]). The leaching reagents are usually used to dissolve waste cathode materials obtained after pretreatment of spent LIBs, and the reductants are used for the reduction of metal ions with high valences (e.g. Co(III), Mn(IV)). With combined reaction of leaching agents and reductants, waste cathode materials can be effectively dissolved or recovered after leaching.

In order to dissolve or recover valuable metals from spent LIBs, different chemical reagents are adopted during the leaching processes. The leaching agents used include mineral acids and organic acids. This chapter will focus on comprehensive summary and analysis of current of above reagents during the leaching reactions.

2.1.1 Mineral Acid Based Leaching

As strong acids, the minerals acids are among the most welcomed reagents during leaching for its excellent performances during leaching in terms of high leaching rate, easily available, low price etc. Table 1 lists a summary review of mineral acids as leaching reagents applied in current studies, including HCl, H_2SO_4, HNO_3 and H_3PO_4. It can be concluded from this table that almost all of valuable metals (e.g. Co, Li, Mn, Ni) can be leached from different kinds of waste cathode materials (e.g. LCO, NCA) using HCl as leaching agent under relatively high acid concentrations (at a range from 1.75 to 4 M), relatively high reaction temperatures over 50 °C, retention times more than 1 h and relatively low pulp density less than 50 g/L. The HCl can itself be oxidized to Cl_2 and treated as reductant during leaching reactions. Therefore, no extra reductants are required during leaching. Through sound results can be obtained using HCl as leaching reagent, the adverse impacts involved with the HCl leaching process may be worse than its positive ones. The first one is the negative environmental effect caused by poisonous gases of Cl_2 generated during leaching reactions, which should be seriously taken into consideration during leaching processes. Besides, it is well known that the high acidity of HCl combined with corrosivity of Cl^- towards equipment is also a serious issue during the leaching processes. Last but not least, excellent dissolvability of HCl may lead to a poor selectivity towards target metal values during the recycling processes, and the rigorous leaching condition adopted (e.g. high reaction temperature and acid concentration) may further increase the subsequent difficulty during the recycling processes.

For the leaching of valuable metals using H_2SO_4, however, it can be concluded from Table 1 that both high and low concentrations of acid (e.g. high as 3 M, low as 0.3 M) are adopted during the leaching of different kinds of waste cathode materials. Under higher acid concentrations over 2 M, almost all metal values (leaching efficiency over 90%) can be leached under specific leaching conditions, which is similar with the leaching results using HCl as leaching reagent. However, the leaching efficiency is not preferable as demanded under lower acid concentrations (e.g. 1 M) and only about 70% Co, 30% Mn and 70% Ni can be leached from mixed

Table 1 Summary of inorganic acid leaching of valuable metals from spent lithium batteries

Acid	Sample	Optimized parameter	Leaching efficiency	Related chemical reaction	References
HCl	NCM	1.75 M, 50 °C, 2 h, 20% (w/v)	99% Co/Mn	$3LiNi_{1/3}Co_{1/3}Mn_{1/3}O_2 + 12HCl \rightarrow 3LiCl + CoCl_2 + NiCl_2 + MnCl_2 + 6H_2O + 1.5Cl_2$	[56]
	LCO	2 M, 60–80 °C, 1.5 h	100% Co/Li	$4HCl + LiCoO_2 \rightarrow LiCl + CoCl_2 + 2H_2O + 0.5Cl_2$	[57]
	LCO	4 M, 80 °C, 2 h,	97% Li, 99% Co		[58]
	LCO	4 M, 80 °C, 1 h, 10 g/L	99% Li/Co		[47]
	Mixed cathodes	4 M, 80 °C, 1 h, 20 g/L	99% Li/Co/Mn/Ni	–	[28]
	NCA	4 M, 90 °C, 18 h, 50 g/L	100% Li/Co/Mn/Al	–	[59]
H_2SO_4	Mixed cathodes	1 M, 95 °C, 4 h, 50 g/L	93% Li, 66% Co, 50% Mn, 96% Ni	$12LiNi_{1/3}Co_{1/3}Mn_{1/3}O_2 + 18H_2SO_4 \rightarrow 6Li_2SO_4 + 4NiSO_4 + 4CoSO_4 + 4MnSO_4 + 18H_2O + 3O_2$	[60]
	Mixed batteries	1 M, 25 °C, 0.5 h, 180 g/L	70% Co, 100% Zn, 30% Mn, 70% Ni		[55]
	LFP	2.5 M, 60 °C, 4 h, 100 g/L	97% Li, 0.027% Fe	–	[61]
	LCO	3 M, 70 °C, 6 h, 200 g/L	98% Li/Co	$3H_2SO_4 + 2LiCoO_2 + 2H_2O_2 \rightarrow Li_2SO_4 + 5H_2O + 1.5O_2 + 2CoSO_4$	[62]
	LCO	2 M, 2 vol.% H_2O_2, 60 °C, 2 h, 33 g/L	88% Li, 96% Co		[29]
	LCO	2 M, 5 vol.% H_2O_2, 75 °C, 1 h, 100 g/L	99% Li, 70% Co		[63]
	LCO	2 M, 5 vol.% H_2O_2, 80 °C, 1 h, 50 g/L	99% Li/Co		[18]

(continued)

Table 1 (continued)

Acid	Sample	Optimized parameter	Leaching efficiency	Related chemical reaction	References
	LCO	2 M, 5 vol.% H_2O_2, 75 °C, 0.5 h, 100 g/L	94% Li, 93% Co		[64]
	LCO	2 M, 6 vol.% H_2O_2, 60 °C, 1 h, 100 g/L	99% Co		[65]
	LCO	4% (v/v), 1 vol.% H_2O_2, 40 °C, 1 h, 33.3 g/L	100% Li, 97% Co		[66]
	LCO	6% (v/v), 1 vol.% H_2O_2, 65 °C, 1 h, 33.3 g/L	95% Li, 80% Co		[67]
	LCO& NiMH	3 M, 3 wt% H_2O_2, 70 °C, 5 h, 66.7 g/L	90% Co/Ni	–	[68]
	NCM	1 M, 1 vol.% H_2O_2, 40 °C, 1 h, 40 g/L	100% Li/Co/Mn/Ni	$6LiNi_{1/3}Co_{1/3}Mn_{1/3}O_2 + 9H_2SO_4 + 3H_2O_2 \rightarrow 3Li_2SO_4 + 2NiSO_4 + 2CoSO_4 + 2MnSO_4 + 12H_2O + 3O_2$	[69]
	NCM	2 M, 5 vol.% H_2O_2, 60 °C, 2 h, 100 g/L	99% Li/Co/Mn/Ni		[70]
	LEP	0.3 M, 60 °C, 2 h, $n(H_2O_2/H_2SO_4{:}Li) = 2.07/0.57$	97% Li, 0.027% Fe, 1.95% P	$2LiFePO_4 + H_2SO_4 + H_2O_2 \rightarrow Li_2SO_4 + 2FePO_4 + 2H_2O$	[46]
HNO_3	LCO	1 M,1.7 vol.% H_2O_2, 75 °C, 1 h, 20 g/L	95% Li/Co	$3HNO_3 + LiCoO_2(s) + 3H_2O_2 \rightarrow LiNO_3 + Co(NO_3)_2 + 4.5H_2O + 0.75O_2(g)$	[71]
	LCO	1 M, 1 vol.% H_2O_2, 80 °C, 1 h, 20 g/L	100% Li/Co		[72]

(continued)

Table 1 (continued)

Acid	Sample	Optimized parameter	Leaching efficiency	Related chemical reaction	References
H_3PO_4	LCO	2% (v/v), 2 vol.% H_2O_2, 90 °C, 0.5 h, 8 g/L	>95% Li/Co	$3LiCoO_2 + 3H_3PO_4 + 6H_2O_2 \rightarrow 10.5H_2O + Co(H_2PO_4)_2 + LiH_2PO_4 + 0.75\ O_2$	[51]
	LCO	1.5 M, 0.02 M glucose, 90 °C, 2 h, 20 g/L	100% Li, 98% Co	–	[73]
	LCO	0.7 M,4 vol.% H_2O_2, 40 °C, 1 h, 20 mL/g	Li/Co > 99	–	[43]

cathodes of spent LIBs as reported by Tanong et al. [55]. Besides, it can be also concluded that the reaction temperatures are lower than that of HCl leaching system and extra reductant (mainly H_2O_2) are needed during the leaching process. Similar with HCl leaching system, all most all valuable metals are dissolved in the H_2SO_4 leaching system and with relatively poor selectivity, except during the leaching of LFP. For the leaching of waste LFP, Li can be completely dissolved while only a very small proportion of Fe can be leached without the addition of reductant of H_2O_2. The H_2SO_4 leaching system can be applicable to almost all kinds of waste cathode materials with relatively high recovery efficiency. However, the high corrosivity and low selectivity listed above should be taken into consideration and further improvement or alternative should be investigated to overcome these issues.

Then HNO_3 and H_3PO_4 leaching systems are established as alternatives to achieve excellent leaching efficiency and prevent the rigorous leaching conditions. It can be discovered in the HNO_3 leaching system that only a moderate acid concentration of 1 M is required for the nearly complete dissolution of waste cathode materials. And the dosage of reductant (i.e. H_2O_2) can be also greatly reduced and only 1 vol.% or 1.7 vol.% H_2O_2 is required in the HNO_3 leaching system. Besides, the leaching solution containing valuable metals can be directly used as raw materials for preparation of new cathode materials after purification. For H_3PO_4 leaching system, it can be discovered that a relatively high acid concentration (e.g. 1.5 M, 2% (v/v)) may result in complete dissolution of Co and Li in the leaching solution a relatively low acid concentration (0.7 M) will lead to the selective leaching of Li and the direct precipitation of Co (as the form of phosphate—$Co_3(PO_4)_2$), indicating that Co and Li can be simultaneously separated and recovered as the form of Li^+ enriched solution and Co^{2+} enriched precipitate. This coupled reaction is also an absolutely novel route to recover metal value from waste cathode materials. However, it can be also found from HNO_3 and H_3PO_4 leaching systems that nitric acidic and phosphoric acidic media are only applicable in relatively simple waste stream (e.g. waste cathode material of $LiCoO_2$) and more complicated waste streams (e.g. waste $LiNi_xCo_yMn_zO_2$) may result in the inapplicability of these acids. Thus, it is necessary to explore recycling routes with wide applicability towards the already complicated waste stream.

In conclusion, mineral acids are widely applied during the leaching processes with advantages of high leaching efficiency, easy availability, relatively low price and so on. According to results from a comparison of different acids listed in Table 1, HCl and H_2SO_4 present the highest leaching efficiencies and they are also capable for the complete dissolution of different metal values from different kinds of waste cathode materials of spent LIBs. Besides, it can be also discovered from Table 1 that HNO_3 and H_3PO_4 also indicate a perfect performances during the leaching of some simple waste cathode materials, such as $LiCoO_2$, and the consumption of acids and reductants can be also greatly reduced in HNO_3 and H_3PO_4 leaching systems, indicating that the two acids can be used for the leaching of a relatively simple waste stream. Besides, it can be discovered that only H_3PO_4 is a relatively weaker acid with potentially characteristics with a combination of dissolving and precipitating in a single leaching step, in which metal values in waste cathode materials can be recovered as different forms after leaching. Based on above analysis, it is necessary to utilize both dissolution characteristics of strong mineral acids like HCl, H_2SO_4 and HNO_3 and the selectivity characteristic like H_3PO_4 during the leaching processes to achieve satisfied recovery results.

2.1.2 Organic Acid Based Leaching

Mineral acids are sophisticated leaching reagents during leaching processes with excellent recovery efficiencies. Meanwhile, an increasing attention has been transferred to the application of different organic acids during the leaching processes due to the serious secondary pollution caused by strong mineral acids, such as poisonous gases, excessive waste acid and other unreacted solid wastes, which a further invest will be required for the alleviation and elimination of the potentially hazardous materials. Therefore, greener organic acids with low poisonousness and biodegradable characteristic become more and more popular during the leaching process. It can be concluded from Table 2 that about more than 14 kinds of organic acids, to our knowledge, are employed as leaching reagents for the leaching of different metal values from waste cathode materials of spent LIBs. Here, in this chapter, we select several typical organic acid for the introduction and analysis of current leaching process.

Citric acid: As a typical tribasic acid, one molar citric acid (H_3Cit) can potentially release three molars of H^+ and Cit^{3-} (including H_2Cit^- and $HCit^{2-}$) can be treated as chelating reagent with heavy metals leached out in the acidic medium. Therefore, the chelation of different metal ions with Cit^{3-}, H_2Cit^- and $HCit^{2-}$) in the leaching solution may promote the leaching reactions forwards, even under a low acid concentration and mild leaching conditions. It can be discovered from Table 2 that the maximum acid concentration is 2 M for leaching of valuable metals using citric acid, under which a satisfied leaching efficiencies can be obtained. It can be also discovered that the dosage of reductant can be reduced compared to reductive leaching using strong mineral acids, such as only about 1 vol.% H_2O_2 and 0.02 M ascorbic acid are used. The citric acid can be not only used for the leaching of waste LCO materials, but also capable for the leaching of metal values of waste NCM materials. Moreover, an extreme low acid concentration of 0.1 M H_3Cit was adopted by Nayaka et al. for the leaching of Co and Li from waste LCO materials using ascorbic acid as reductant, and Li can be completely dissolved and 80% of Co can be leached in the mixed acidic medium [74]. As a common employed weak acid, citric acid may demonstrate the following advantages: (1) Citric acid can be itself used as both acid and chelating reagent, which can greatly improve the leaching efficiency; (2) Relatively high leaching efficiencies can be also achieved by using citric acid with lower acid concentration and mild leaching conditions compared with strong mineral acids; (3) Citric acid can be an environmentally friendly acid without adverse impacts on the eco-system and the unreacted acid can be easily degraded during and after leaching. However, the usage of citric acid may also get into troubles in aspects of: (1) enhanced invest caused by the high price of acid, (2) The difficulty in subsequent separation and recovery of metal values from leaching solutions caused the

Table 2 Summary of organic acid leaching of valuable metals from spent lithium batteries

Acid	Sample	Optimized condition	Leaching efficiency	Related chemical reaction	References
Citric acid	LCO	1.25 M, 1 vol.% H_2O_2, 90 °C, 0.5 h, 20 g/L	100% Li, > 90% Co	$6H_3Cit + 2LiCoO_2 + H_2O_2 \rightarrow 2Li^+ + 6H_2Cit^- + 2Co^{2+} + 4H_2O + O_2$ $6H_2Cit^- + 2LiCoO_2 + H_2O_2 \rightarrow 2Li^+ + 2Co^{2+} + 6HCit^{2-} + 4H_2O + O_2$ $6HCit^{2-} + 2LiCoO_2 + H_2O_2 \rightarrow 2Li^+ + 2Co^{2+} + 6Cit^{3-} + 4H_2O + O_2$	[31]
	LCO	2 M, 1.25 vol.% H_2O_2, 60 °C, 2 h, 30 g/L	92% Li, 81% Co		[31]
	LCO	2 M, 0.6 g/g H_2O_2, 70 °C, 80 min, 50 g/L	99% Li, 98% Co		[21]
	NCM	2 M, 2 vol.% H_2O_2, 80 °C, 1.5 h, 33.3 g/L	99% Li, 95% Co, 94% Mn, 97% Ni	$2LiNi_{1/3}Co_{1/3}Mn_{1/3}O_2 + 2H_3Cit + H_2O_2 \rightarrow 2Li^+ + 2/3Mn^{2+} + 2/3Co^{2+} + 2/3Mn^{2+} + 4Cit^{3-} + 4H_2O + O_2$	[79]
	LCO	0.1 M, 0.02 M ascorbic acid, 80 °C, 6 h, 10 g/L	100% Li, 80% Co	–	[74]
Ascorbic acid	LCO	1.25 M, 70 °C, 20 min, 25 g/L	99% Li, 95% Co	$4C_6H_8O_6 + 2LiCoO_2 \rightarrow C_6H_6O_6 + C_6H_6O_6Li_2 + 2C_6H_6O_6Co + 4H_2O$	[75]

(continued)

Table 2 (continued)

Acid	Sample	Optimized condition	Leaching efficiency	Related chemical reaction	References
Acetic acid	NCM	3.5 M, 4 vol.% H_2O_2, 60 °C, 1 h, 40 g/L	100% Li, 94% Co, 96% Mn, 93% Ni	$6LiNi_{1/3}Co_{1/3}Mn_{1/3}O_2 + 18CH_3COOH + 3H_2O_2 \rightarrow 2(CH_3COO)_2Ni + 2(CH_3COO)_2Co + 2(CH_3COO)_2Mn + 6CH_3COOLi + 12H_2O + 3O_2$	[80]
	LCMO	3 M, 7.5 vol.% H_2O_2, 70 °C, 40 min, 20 g/L	100% Li, 99% Co, 100% Mn	$Li_2CoMn_3O_8 + 10CH_3COOH + 10H_2O_2 \rightarrow (CH_3COO)_2Co + 3(CH_3COO)_2Mn + 2CH_3COOLi + 8H_2O + 3O_2$	[81]
	LEP	0.8 M, 6 vol.% H_2O_2, 50 °C, 0.5 h, 120 g/L	95% Li, 94% Fe	$LiFePO_4 + CH_3COOH + 0.5H_2O_2 \rightarrow FePO_4 + CH_3COOLi + H_2O$	[82]
Malic acid	LCO	1.5 M, 2 vol.% H_2O_2, 90 °C, 40 min, 20 g/L	100% Li, 90% Co	$6H_2C_4H_4O_5 + 2LiCoO_2 + H_2O_2 \rightarrow 2Li^+ + 2Co^{2+} + 6HC_4H_4O_5^- + 4H_2O + O_2$ $6HC_4H_4O_5^- + 2LiCoO_2 + H_2O_2 \rightarrow 2Li^+ + 2Co^{2+} + 6C_4H_4O_5^- + 4H_2O + O_2$	[83]

(continued)

Table 2 (continued)

Acid	Sample	Optimized condition	Leaching efficiency	Related chemical reaction	References
	NCM	1.2 M, 1.5 vol.% H_2O_2, 80 °C, 30 min, 40 g/L	99% Li, 94% Co, 96% Mn, 95% Ni	$6H_2C_4H_4O_5 + 3LiNi_{1/3}Co_{1/3}Mn_{1/3}O_2 + H_2O_2 \rightarrow 3Li^+ + Co^{2+} + Mn^{2+} + Ni^{2+} + 6HC_4H_4O_5^- + 4H_2O + 2O_2$ $6HC_4H_4O_5^- + 3LiNi_{1/3}Co_{1/3}Mn_{1/3}O_2 + H_2O_2 \rightarrow 3Li^+ + Co^{2+} + 6C_4H_4O_5^- + 4H_2O + 2O_2$	[84]
Maleic acid	NCM	2 M, 2 mL H_2O_2, 70 °C, 1 h, 20 g/L	98% Li/Co/Mn/Ni	–	[85]
Oxalic acid	LCO	1 M, 80 °C, 2 h, 50 g/L	98% Li/Co	$7H_2C_2O_4 + 2LiCoO_2 \rightarrow 2LiHC_2O_4 + 2Co(HC_2O_4)_2 + 4H_2O + 2CO_2$	[76]
	LCO	1 M, 95 °C, 2.5 h, 15 g/L	98% Li, 97% Co	$4H_2C_2O_4 + 2LiCoO_2 \rightarrow Li_2C_2O_4 + 2CoC_2O_4 + 4H_2O + 2CO_2$	[77]
Tartaric acid	LCO	0.6 M, 3 vol.% H_2O_2, 80 °C, 30 min, 30 g/L	97% Li, 98% Co	$2LiCoO_2 + 3C_4H_6O_6 + H_2O_2 \rightarrow 2C_4H_4O_6Co + C_4H_4O_6Li_2 + 4H_2O + O_2$	[78]

(continued)

Table 2 (continued)

Acid	Sample	Optimized condition	Leaching efficiency	Related chemical reaction	References
	NCM	2 M, 4 vol.% H_2O_2, 70 °C, 30 min, 17 g/L	99% Li/Co/Mn/Ni	$10LiNi_{0.5}Co_{0.2}Mn_{0.3}O_2 + 15C_4H_6O_6 + 5H_2O_2 \rightarrow 2C_4H_4O_6Co + 5C_4H_4O_6Li_2 + 5C_4H_4O_6Ni + 3C_4H_4O_6Mn + 20H_2O + 5O_2$	[26]
Formic acid	NCM	2 M, 6 vol.% H_2O_2, 60 °C, 30 min, 17 g/L	100% Li, 85% Co/Mn/Ni	$18HCOOH + 6LiNi_{1/3}Co_{1/3}Mn_{1/3}O_2 + 3H_2O_2 \rightarrow 2C_2H_2NiO_4 + 2C_2H_2CoO_4 + 2C_2H_2MnO_4 + 6CHLiO_2 + 12H_2O + 3O_2$	[86]
Succinic acid	LCO	1.5 M, 4 vol.% H_2O_2, 70 °C, 40 min, 15 g/L	96% Li, 100% Co	$3C_4H_6O_4 + 2LiCoO_2 + H_2O_2 \rightarrow C_4H_4O_4Li_2 + 2C_4H_4O_4Co + 4H_2O + O_2$	[50]
Trichloroacetic acid	NCM	3 M, 4 vol.% H_2O_2, 60 °C, 30 min, 50 g/L	100% Li, 92% Co, 90% Mn, 93% Ni	$3LiNi_{1/3}Co_{1/3}Mn_{1/3}O_2 + 9CCl_3COO^- + 9H^+ \rightarrow 3Li(CCl_3COO) + Ni(CCl_3COO)_2 + Co(CCl_3COO)_2 + Mn(CCl_3COO)_2 + 4.5H_2O + 0.75O_2$	[87]

(continued)

Table 2 (continued)

Acid	Sample	Optimized condition	Leaching efficiency	Related chemical reaction	References
Lactic acid	NCM	1.5 M, 0.5 vol.% H_2O_2, 70 °C, 20 min, 20 g/L	98% Li, 99% Co, 98% Mn, 98% Ni	$3LiNi_{1/3}Co_{1/3}Mn_{1/3}O_2 + 9C_3H_6O_3 + 0.5H_2O_2 \rightarrow 3 C_3H_5O_3Li + (C_3H_5O_3)_2Ni + (C_3H_5O_3)_2Co + (C_3H_5O_3)_2Mn + H_2O + O_2$	[27]
Glycine	LCO	0.5 M, 0.02 M ascorbic acid, 80 °C, 6 h, 2 g/L	91% Co	–	[88]
Iminodiacetic acid	LCO	0.5 M, 0.02 M ascorbic acid, 80 °C, 6 h, 2 g/L	99% Li, 91% Co	$3C_4H_7NO_4 + 2LiCoO_2 + H_2O_2 \rightarrow C_4H_5O_4NLi_2 + 2C_4H_5O_4NCo + 4H_2O + O_2$	[89]
Aspartic acid	LCO	1.5 M, 4 vol.% H_2O_2, 90 °C, 2 h, 10 g/L	100% Li, 90% Co	$3C_4H_7NO_4 + 2LiCoO_2 + H_2O_2 \rightarrow C_4H_5O_4NLi_2 + 2C_4H_5O_4NCo + 4H_2O + O_2$	[90]

strong chelation and (3) The issue of waste water treatment caused by large amount of high level of COD solution after leaching.

Ascorbic acid: The reason we select this acid as a typical one is that the ascorbic acid can be both used as leaching reagent and reductant during leaching, while all other acids can be just used as leaching reagent without reducing characteristics. As reported by Li et al. [75], ascorbic acid is selected as leaching and reducing reagent to improve the recovery efficiency of Co and Li after the pretreatment procedures of ultrasonic washing for peeling off Al foils and calcination for removal of PVDF binder. About 95% Co and 99% Li can be leached out under the optimized experimental conditions of acid concentration—1.25 M, reaction temperature—70 °C, retention time—20 min and slurry density—25 gL^{-1}. It can be also discovered from Table 2 that the leaching reaction can be finished in a short time of 20 min and only mild leaching conditions are required during leaching, which can greatly reduce the energy and chemical consumption. However, the leaching of metal values using ascorbic acid may also encounter some obvious issues. One point in this case is the easy oxidation property of ascorbic acid in air atmosphere, in which it may react with oxygen in the air to degenerative acid and directly affect its leaching ability. Therefore, extra protection of inert gases may be required during the leaching processes, which may increase the cost during leaching. Besides, the separation and recovered of different metals from the leaching solution is also a difficult issue because ascorbic acid is a strong chelating reagent and the common chemicals may be invalid during the subsequent separation and recovery processes. Finally, we found in that some unknown precipitates are formed after staying the leaching solution for several days, which may be caused by the combination of different functional groups and the formation of precipitates as macromolecular compounds.

Oxalic acid: As another kind of typical organic acid, oxalic acid can be also used as leaching reagent for the dissolving of waste LCO materials from spent LIBs [76, 77]. Sun and Qiu reported the possibility for the selective leaching of Li and direct precipitation of Co as CoC_2O_4 using 1.0 M oxalic acid as leaching agent under experimental conditions of reaction temperature—80 °C, slurry density—50 gL^{-1} and retention time—120 min [76]. It is found that oxalic acid can be applied as both leaching and precipitating reagent, and a reaction efficiency over 98% of $LiCoO_2$ can be achieved with this simple hydrometallurgical process. Afterwards, Zeng et al. further investigated this leaching process using oxalic acid as leaching and precipitating reagent under the optimized leaching conditions of retention time—150 min, heating temperature—95 °C, solid-liquid ratio—15 gL^{-1} and rotation rate—400 rpm, and about 98% Li and 97% Co can be respectively recovered as Li^+ enriched solution and Co enriched precipitate in a single leaching step [77]. It can be concluded that both the above studies prove the possibility using oxalic acid as leaching and precipitating reagent and sound results can be obtained under optimized leaching conditions. However, it can be also discovered that the selective leaching and precipitating can be only applicable for relatively simple waste stream of LCO because only Co can be effectively precipitated using $H_2C_2O_4$ as the precipitant and other metals (e.g.

Mn, Fe) cannot be applicable in the oxalic acidic medium. Thus, sustainable leaching process with wider application will be still required for the efficient recovery of various metal values from spent LIBs.

Tartaric acid: Similar with oxalic acid, tartaric acid is also a dibasic acid with weaker acidity than that of oxalic acid. But with a higher molecular weight, tartaric acid shows a totally different characteristics. As reflected in Table 2, a higher acid concentration will result in a nearly complete dissolution of waste cathode materials while a short-cut process can be achieved under lower acid concentration for the simultaneous recovery of Co and Li from waste cathode materials of spent LIBs. As reported by He et al., a hydrometallurgical process involved with L-tartaric acid leaching for the sustainable recovery of Ni, Co, Mn and Li from spent LIBs was proposed [26]. About 99% Mn, 99% Li, 99% Co and 99% for Ni can be leached under the optimized leaching conditions of reductant dosage—4 vol.% H_2O_2, acid concentration—2 M L-tartaric acid, pulp density—17 g/L, leaching temperature—70 °C and duration—30 min. This proposed recycling process can be simple, efficient, and environmentally-friendly route for the efficient dissolution of valuable metals in the tartaric acidic medium. Similarly, a short-cut process was proposed in our latter study using mild tartaric acid (0.6 M) as leaching and chelating reagent for the selective precipitating of Co from waste cathode materials of spent LIBs [78]. About 98% of Co and 97% of Li can be simultaneously recovered as Co^{2+} enriched precipitate and Li^+ enriched solution after a single leaching step under the optimized experimental conditions of reaction temperature—80 °C, duration time—30 min, slurry density—30 mL/g and reducatant dosage—3 vol.% H_2O_2. Therefore, the tartaric acid also demonstrates a double-planedness during leaching processes and if we want to completely dissolve valuable metals in acidic solution, a higher acid concentration should be chosen; if we want to separate and extract some target metals, a mild acid concentration should be then a wise choice.

The above four kinds of organic acids are presented as representatives for acids with specific features during leaching. For other organic acids, it can be also concluded from Table 2 that almost all of those acids demonstrate the similar leaching ability as those of mineral acids. Besides, it can be also discovered that lower acid concentration (usually less than 2 M) is required during the leaching processes compared with the leaching processes using mineral acids (usually higher than 2 M). Finally, reaction temperature and duration time can be also greatly reduced when compared organic acids with the mineral acids. Therefore, the usage of organic acids for the leaching process is an obvious progress, not only for the consideration of environmental protection, but also for the sake of enhanced leaching efficiency and reduced energy consumption. However, issues caused by the strong chelation of organic acids with metal ions should be also taken into consideration for the following metal separation and recovery processes.

2.2 Bioleaching Process

Recently, an increasing attention has been paid on the leaching of valuable metals from spent LIBs using bioleaching method. Bioleaching belongs to one of the hydrometallurgical step during metal recycling processes. Different from traditional leaching process using different chemicals as leaching reagents, bacteria, microorganism and their microbial secondary metabolites are usually used during the bioleaching process. Current processes involved with recycling of valuable metals from spent LIBs mainly concentrate on direct employing of microorganism or indirect utilizing of microbial metabolites during the leaching process. Therefore, this chapter will focus on the above two aspects (direct bioleaching process and indirect bioleaching process) for the brief introduction and discussion of current bioleaching processes.

Direct bioleaching process: The direct bioleaching process usually employs different kinds of bacteria for the leaching of valuable metals from solid spent LIBs under controlled leaching conditions. For example, Xin et al. investigated the recovery of valuable metals of Li, Co, Mn and Ni from different kinds of waste cathodes—$LiFePO_4$, $LiMn_2O_4$ and $LiNi_xCo_yMn_{1-x-y}O_2$ of spent EV LIBs by bioleaching method at a pulp density of 1% in different leaching systems [14]. According to their results, about 95% of 4 kinds of valuable metals of Li, Ni, Co and Mn can be leached under the optimized leaching conditions, indicating that the cheap autotrophic bioleaching applied in this study can be effective candidate for the recovery of valuable metals from spent EV LIBs. Besides, it can be discovered that the maximum recovery efficiency of Li occurred in the sulfur bioleaching system, indicating that the recovery of Li can be attributed to acid solution by the biogenic H_2SO_4, and the mixed energy source-mixed culture bioleaching system presents the highest leaching yield for Co, Ni and Mn than other bioleaching systems, suggesting that these transition metals can be mobilized by a combined effect of Fe^{2+} reduction and acid dissolution. Another example for this direct bioleaching process is the application of different kinds of bacteria for the dissolution of valuable metals from spent LIBs as reported by Biswal et al. [91]. In this study, *Aspergillus niger* strains of MM1 and SG1 and *Acidithiobacillus thiooxidans* of 80,191 were adopted for during the bioleaching process, and fungal leaching for Co from solutions of sodium sulfide, sodium hydroxide and sodium oxalate and then sodium carbonate solution for Li was also detailed investigated. It is discovered that a relatively high leaching efficiencies of 82 and 100% for Co and Li can be obtained in strain of MM1, while only 22% Co and 66% Li are solubilized in bioleaching system of strain 80,191. Afterwards, high precipitation rates can be obtained for Co (cobalt sulfide—100%, cobalt hydroxide—100% and cobalt oxalate—88%) and Li (lithium carbonate — 73.6%). In summary, it can be concluded that the direct bioleaching method can be an effective route for the recovery of valuable metals from spent LIBs and it also demonstrate with advantages of mild leaching conditions, chemical consumption saving and environmentally friendly process. However, this direct bioleaching process may be also discouraged by several issues in terms of long bioleaching time, restricted application in the already complicated waste stream, difficulties during the cultivation and screening of high active bacteria.

Indirect bioleaching process: Different from above direct bioleaching process, the indirect bioleaching route usually utilizes the metabolites generated by different kinds of microorganisms for the efficient extraction of valuable metals from spent LIBs. This indirect bioleaching process always avoids direct contact of microorganisms with waste streams. For example, the application of indirect and non-contact bioleaching in biogenic ferric iron and sulfuric acid media was applied by Boxall et al. for the dissolution of valuable metals from spent LIBs [92]. They found that the leaching efficiencies for different metals can be greatly improved by multiple sequential leaching stages (4 × 1 h), rather than a single leach stage at ambient temperature (with only a low yields less than 10%). And highest leaching yields of 53.2% Co, 60.0% Li, 48.7% Ni, 81.8% Mn and 74.4% Cu can be achieved by biogenic ferric leaching augmented with 100 mM H_2SO_4, indicating that the extra addition of H_2SO_4 can facilitate the bioleaching process which may be attributed for the supply of additional sulfur source in the bioleaching system. It can be concluded that application of bioreagents generated from microbial metabolites is viable and this process can be also a more environmentally benign alternative than these traditional mineral processing, which can be further improved after the appropriate pre-treatment of spent LIBs. Moreover, a functional exploration of extracellular polymeric substances (EPS) was conducted by Wang et al. during the bioleaching of obsolete electric vehicle only constituted of $LiNi_xCo_yMn_{1-x-y}O_2$ Li-ion batteries [93]. Based on their discoveries, EPS plays an important role during the attachment of cells to cathodes, and then aggregates can be formed (i.e. cell-EPS-cathode), which is variation with the electrical and surface properties of aggregates, concentration of both Fe^{2+} and Fe^{3+} surrounding the aggregates, electron transfer inside the aggregates and metals released from the aggregates. Both Fe^{3+} and Fe^{2+} ions will be adsorbed by EPS to ensure the Fe^{2+}/Fe^{3+} cycle inside the aggregates, which may lead to a stronger reductive attack on the high valence state of transition metals (e.g. Co(III), Mn(IV)) as a contact reductive mechanism. Finally, they found that the retention/addition of EPS will increase electronic potential and reduce electronic resistance to promote corrosion electric current, thereby boosting the electron transfer and metal dissolution. Despite the indirect bioleaching method is theoretically feasible as proved by previous studies, it can be discovered that leaching efficiencies for valuable metals are in a relatively low level. Therefore, it is still necessary to take advantages of bioleaching and develop new methods to improve the yields during the recycling processes.

3 Separation and Recovery Processes

After the above leaching process, different metals in spent LIBs, including valuable metals, are dissolved in leaching solutions or directly recovered as precipitate form. Solid residues, including formed precipitate compounds, can be removed or recovered after filtration. And the filtrate is the leaching solution and different metals are dissolved in this leaching liquor. The main technologies for the separation and extraction of different metals from above leaching solution mainly include chemical precipitation, solvent extraction, adsorption, ion exchange and membrane separation etc. In this chapter, we will mainly focus on the reviewing and discussing of these most sophisticated processes for the separation and recovery of different metals from the leaching solutions.

3.1 Chemical Precipitation

As one of the most traditional separation technologies, the precipitating of different metals in leaching solutions is efficient method during the purification and recycling processes. Usually, a variety of chemical reagents, called as precipitants, are employed for the effective purification and separation of different metals based on their chemical properties. Among them, hydroxides (e.g. NaOH), carbonates (e.g. Na_2CO_3), strong oxidants ($KMnO_4$), sulfides (e.g. $(NH_4)_2S$), phosphates (e.g. Na_3PO_4), oxalates (e.g. $Na_2C_2O_4$), chelating precipitants (e.g. dimethylglyoxime, $C_4H_8N_2O_2$) etc. are the mostly employed chemical reagents during the precipitating reactions. In this chapter, several precipitants are selected as the typical examples for the introduction and analysis current chemical precipitation processes.

Sodium hydroxide: Sodium hydroxide (NaOH) is the most widely used precipitant for both the purification and metal precipitation processes. It can be concluded from Table 3 that valuable metals of Ni, Co and Mn can be sequentially precipitated under different pHs of 8, 10 and 12, indicating that valuable metals in leaching solution can be selectively recovered by controlling equilibrium pH of relevant precipitation reactions. Besides, it can be also discovered that NaOH can be also employed for the removal of impurity ions (i.e. Fe, Al, Cu) from the obtained leaching solutions under relatively low pH (e.g. pH = 5.3). When the pH of leaching solutions is relatively high, low value-added metals of Mn/Ca/Mg can be also removed by adding NaOH solution to a pH of 12. Finally, it can be also discovered from Table 3 that NaOH solution is can be also employed as precipitant for the co-precipitation of valuable metals as precursor materials of $Ni_xCo_yMn_z(OH)_2$, which can be used for the direct re-preparation of new cathode materials from leaching solution. However, extra inert gases should be employed for the prevention of oxidation process when transition metals (Ni, Co and Mn) are exposed to the atmosphere for the transition metals can be easily oxidized into their high valences with their exposure to oxidant, such as O_2 in the air.

Sodium carbonate: Similar with NaOH, sodium carbonate (Na_2CO_3) are also widely used as precipitant for the precipitation of valuable metals from the leaching solutions, and Ni, Co and Mn demonstrate similar optimized pH (around 9–10) during the precipitating reactions. However, the recovery of Li is much more difficult than the above transition metals (Ni, Co and Mn) when using Na_2CO_3 as precipitant and it is necessary to use saturated Na_2CO_3 under a boiling condition for the

Table 3 A summary of the chemical precipitating processes

Reagent	Element	Precipitate	pH	References
NaOH	Mn	$Mn(OH)_2$	12	[28]
	Ni	$Ni(OH)_2$	8	[28, 59]
	Co	$Co(OH)_2$	10	[28, 94]
	Fe/Al/Cu	$Fe(OH)_3/Al(OH)_3/Cu(OH)_2$	5.3	[13, 30, 58, 95–97]
	Zn/Ni/Co	$Zn(OH)_2/Ni(OH)_2/Co(OH)_2$	7	[95]
	Mn/Ca/Mg	$Mn(OH)_2/Ca(OH)_2/Mg(OH)_2$	12	[95]
	Ni/Co/Mn	$Ni_xCo_yMn_z(OH)_2$	11	[70, 98, 99]
$NaCO_3$	Mn	$MnCO_3$	–	[100, 101]
	Ni	$NiCO_3$	9	[101]
	Co	$CoCO_3$	9–10	[96, 102]
	Li	Li_2CO_3	–	[28, 86, 96, 100, 103–105]
	Ni/Co/Mn	$(Ni_xCo_yMn_z)CO_3$	7.5–8	[106]
$KMnO_4$	Mn	MnO_2	2	[28]
$(NH_4)_2S$	Co	CoS	6	[100]
Na_3PO_4	Li	Li_3PO_4	–	[23, 95, 105]
$H_2C_2O_4/(NH_4)_2C_2O_4$	Co	$CoC_2O_4{\cdot}2H_2O$	1.5,2	[23, 30, 74]
$C_4H_8N_2O_2$	Ni	$NiC_8H_{15}N_4O_4$	8	[13, 28]
$CaCO_3/O_2$	Fe	$FeO{\cdot}OH$	6	[107]
$Ca(OH)_2$	Fe	Fe_2O_3	7	[107]
H_2S	Fe	FeS_2	–	[107]
NaF	Li	LiF	–	[74, 94]
[a]C8H	Co/Mn	–	8	[108]
CaO	Mg	$Mg(OH)_2$	8.6	[104]
$Na_2C_2O_4$	Mg/Ca	CaC_2O_4/MgC_2O_4	–	[104]

[a]Copper(II)-8-hydroxquinoline system

sufficient precipitating of Li^+ from the leaching solution (with a recovery efficiency around 80%). This phenomenon can be attributed to the active chemical properties of Li over transition metals of Ni, Co and Mn and the relatively low Ksp value of Li_2CO_3 in aqueous phase. Besides, it can be also discovered that Na_2CO_3 can be also employed as an efficient co-precipitating reagent for the recovery of different valuable metals and the re-preparation of precursor- $(Ni_xCo_yMn_z)CO_3$ from leaching solutions. Different from the co-precipitation method reported by using NaOH, it can be found that only relatively low pH (near to neutral aqueous solution, pH = 7.5–8) is required during the carbonation reactions of transition metals (i.e. Ni, Co and Mn) to form the precipitate of $(Ni_xCo_yMn_z)CO_3$ and the prepared precipitate presents good chemical stability in the atmosphere, which can be also used as precursor for the preparation of cathode materials of $LiNi_xCo_yMn_zO_2$. However, it can be discovered that the quality of $(Ni_xCo_yMn_z)CO_3$ precursor may be affected by some impurity ions, especially Li^+ ions, in the leaching solutions because almost all metals can be precipitated by the addition of Na_2CO_3 and the electrochemical performances of $LiNi_xCo_yMn_zO_2$ may be affected by the application of the contaminated $(Ni_xCo_yMn_z)CO_3$ precursor.

Ammonium oxalate: As demonstrated in Table 3, oxalates, including the ammonium oxalate ($(NH_4)_2C_2O_4$), sodium oxalate ($Na_2C_2O_4$) and oxalic acid ($H_2C_2O_4$) can be employed as efficient precipitant for the recovery of Co^{2+} or removal of Mg/Ca impurity ions from the leaching solution. According to previous studies, recovered Co is presented in the form of CoC_2O_4 with relatively high purity. Other metals (e.g. Ni, Mn and Li) cannot be sufficiently precipitated using oxalates as precipitant, which can be attributed to the high Ksp values for other metals. Besides, Li^+ ions in the leaching solution cannot be nearly precipitated as $Li_2C_2O_4$ due to its higher solubility in oxalic acidic medium, which can avoid the contamination of Li during the co-precipitation of other valuable metals of Ni, Co and Mn. For the removal of impurity ions of Mg^{2+} and Ca^{2+} ions, the usage of oxalates is also a wise choice for the low Ksp values of MgC_2O_4 and CaC_2O_4.

Other precipitants: Except the above three precipitants, there also a variety of other kinds of chemicals used for the precipitating or recovery other valuable metals. For example, potassium permanganate ($KMnO_4$) can be used for the selectively precipitation of Mn^{2+} ions in the leaching solutions, and Mn with high valence (e.g. Mn (VII)) in $KMnO_4$ can be usually treated as oxidizing reagent while Mn with low valence (e.g. Mn (II)) in leaching solution can be used as reductant. Manganese dioxide (MnO_2) can be obtained after above comproportionation. Besides, chelating precipitants can be also employed for the selective separation and recovery of target metals in the leaching solution. A case in this point is the usage of dimethylglyoxime (DMG, $C_4H_8N_2O_2$) for the selective precipitation of Ni^{2+} ions from the leaching solution. Li^+ ions in the leaching solution are always chemically active and it is difficult for the sufficient or complete recovery of Li^+ ions from leaching solutions. Therefore, Na_3PO_4 and NaF were employed for the precipitating of Li^+ ions as precipitate form of Li_3PO_4 or LiF due to their low solubility in aqueous phase. Other chemicals, such as $CaCO_3/O_2$, $Ca(OH)_2$, are used for the selective removal of Fe ions. In summary, the usage of different chemicals as precipitant should be determined by different metals and the acidic media and we should adopt the most suitable and effective precipitant for the removal or recovery of the target metals.

3.2 Solvent Extraction

As another typical separation method, the solvent extraction (usually called as liquid-liquid extraction) process is an effective route for the separation and recovery of different metals from leaching solutions. Usually, different organic chemicals with specific functional groups are used as extractants, and these organics (one or more kinds of them) are usually mixed with specific diluent and modification reagents to form the extraction system. The above organics may be saponified in advance with high concentrations of acids or alkalis to form saponification-salt for the sufficient recovery of valuable metals from leaching solutions. As shown in Table 4, different organics are used for the separation and extraction of both valuable metals (e.g. Ni^{2+}, Co^{2+}, Mn^{2+}) and impurity ions (e.g. Fe^{3+}, Al^{3+}) from leaching solutions. This chapter will concentrate on the introduction and discussion of the current solvent extraction processes for the separation and recovery of different metals of Ni, Co, Mn, Li, Cu, Al and Fe from leaching solutions. And we divide these metals into valuable metals—Ni, Co, Mn, Li and Cu and impurities ions—Al, Cu and Fe (here Cu can be treated as both valuable metal and impurity metal).

Solvent extraction of Co: It can be discovered from Table 4 that the separation and recovery of Co from leaching solutions using different kinds of organic extractants is the most sophisticated method. Phosphonic organics of Cyanex272/Cyanex301, PC88A and P507 are typical extractants that are the most widely employed ones for the selective separation and recovery of Co from other metals. For example, Co^{2+} ions can be effectively separated and recovered from leaching solutions by Cyanex272 with an extraction equilibrium pH range of 4.5–5.5 at room temperature and short extraction time less than 20 min. The extraction efficiency for Co is over 90% with relatively high separation factors over other metals of Li, Ni and Mn. Besides, it can be also discovered that other extractants, such as PC88A (also called P507 in China), are also employed for the extraction of Co from the leaching solutions. However, it can be found that, despite a relatively high recovery rate of Co can be obtained during the use of other organic extractans, a certain amount of other metals (e.g. Li, Mn, Ni) can be also co-extracted with Co^{2+} ions. This may aggravate the difficulty for the subsequent scrubbing and purification processes. Therefore, it can be concluded that Cyanex272 would be a better choice for the selective separation and extraction of Co from leaching solutions, especially for the selective recovery of Co from complicated leaching solutions containing other similar metal ions (such as Ni, Mn).

Solvent extraction of Mn: It can be discovered from Table 4 that extractants of D2EHPA and PC88A are the mostly used organic for the separation and recovery of Mn from leaching solutions. As reported in our previous study [79], 20 vol.% D2EHPA with 70% saponification rate was used as extractant for the separation of Mn^{2+} ions from the citric acid leaching solution under extraction conditions of extraction equilibrium pH—4, O/A = 1, contact time—5 min and room temperature, and about 95% Mn can be extracted from leaching solution with a loss of 2% Li. After extraction, the Mn loaded organic can be stripped using 0.2 M H_2SO_4 as scrubbing agent at O/A of 1. In addition, it can be also found that synergistic extraction reagents of PC88A + Cyanex272 can be also used as effective extraction system for the selective separation and recovery of Mn from leaching solution containing Co and Mn ions as reported by Zhao et al. [109]. They found that the separation factor for Mn and Co can reach over 700 ($\beta_{Mn/Co} = 712.7$) under the optimized extraction conditions of room temperature, O/A = 1, t = 0.5 h, and pH = 5. Therefore, it may be concluded that D2EHPA can be an effective phosphonic acid extractant for the sole extraction of Mn from relatively simple leaching solution containing only Mn/Mn and Li, and metals ions with similar chemical properties (e.g. Co^{2+}) contained in leaching solutions may increase the difficulty for the selective separation of Mn, and

Table 4 A summary of the current solvent extraction processes

Element	Organic phase	Optimized condition	Stripping	Yield (%)	References
Co	10 vol.% Cyanex272	pH = 4.8, O/A = 1	–	98% Co	[105]
	10 vol.% PC88A + 5 vol.% TOA	RT, pH = 5.5–6.0, O/A = 1, t = 2 h	3 M H_2SO_4, O/A = 1	90% Co	[110]
Co(Li)	0.9 M PC-88A	RT, pH = 6.7, O/A = 0.85, t = 0.5 h	2 M H_2SO_4, O/A = 5	100% Co	[47]
	0.3 M Cyanex272	T = 50 °C, pH = 4.5, O/A = 1, t = 20 min	3 M H_2SO_4, O/A = 1	68% Co, 39% Li, 98% Al	[111]
	0.72 M Cyanex272	T = 50 °C, pH = 5, O/A = 1, t = 20 min	2.5 M H_2SO_4, O/A = 5	88% Co, 33% Li, 100% Al	[67]
	65% saponified 1.5 M Cyanex272	RT, pH = 5, O/A = 1, t = 10 min	5 M H_2SO_4, O/A = 5	85% Co	[64]
	15% Cyanex272 & 3% isodecanol	RT, pH = 5, O/A = 1, t = 5 min	10% H_2SO_4, O/A = 1	99.9% Co, 7.9% Li	[112]
	65% saponified 48 vol.% Cyanex272 + 5 vol.%TBP	RT, pH = 5, O/A = 2/3, t = 10 min	5 M H_2SO_4, O/A = 5:1	99.9% Co, 35.5% Li	[113]
	10% saponified 1 M Cyanex272	RT, pH = 5.5, O/A = 1, t = 1 min	2 M H_2SO_4, O/A = 1	96% Co	[62]
	0.1 M Cyanex301	RT, pH = 7.2, O/A = 1	2 M H_2SO_4	$\beta_{Co/Li} = 612$	[114]
	20% Cyanex272	pH = 4.8, O/A = 1	–	98% Co	[105]
Co(Ni)	Cyanex272	pH = 5–6, O/A = 1, Cyanex/Co = 4	4 M H_2SO_4, O/A = 1	95% Co	[114]

(continued)

Table 4 (continued)

Element	Organic phase	Optimized condition	Stripping	Yield (%)	References
	10% saponified 1 M Cyanex272	RT, pH = 5.1–5.3, O/A = 1	2 M H_2SO_4, O/A = 1	97% Co	[68]
Co(Ni/Li)	50% saponified 0.4 M Cyanex272	RT, pH = 5.5–6.0, O/A = 2	2 M H_2SO_4, O/A = 11.7:1	95–98% Co, 1% Ni	[65]
	25 wt% P507	RT, pH = 4.15, O/A = 1.5, t = 10 min	3 M H_2SO_4, O/A = 4	95% Co, <5% Ni &Li	[30]
	15% Cyanex272	RT, pH = 5.5–6.0, O/A = 2	80 g/LH_2SO_4, O/A = 8	90% Co	[115]
	Cyanex272 + 10%TBP	RT, pH = 4, O/A = 1, Cyanex/Co = 4	3 M H_2SO_4, O/A = 1:1.5	99.9% Co, 10% Ni & Li	[116]
Co/Ni(Li)	Cyanex272 + 5% v/v TOA	RT, pH = 6.8–7.1, O/A = 1:0.77	H_2SO_4, pH = 5–6	99.6% Ni, 98.7% Co	[117]
Co/Mn/Ni(Li)	40% D2EHPA in kerosene	RT, pH = 3.5, O/A = 1	0.5 M H_2SO_4	85% Ni, 99% Co, 100% Mn	[118]
Co/Mn(Li); Mn(Co)	0.05 M Cyanex272 + 0.05 M PC-88A; Cyanex272 + PC-88A +EDTA	RT, O/A = 1, t = 0.5 h, pH = 5.2; pH = 5	0.01 M H_2SO_4	$\beta_{Mn/Co}$ = 712.7	[109]
Mn	65% saponified D2EHPA	RT, pH = 4, O/A = 1, D2EHPA/Mn = 2	4 M H_2SO_4, O/A = 1	$\beta_{Mn/Co}$ = 16	[114]
Mn(Li)	70% saponified 20 vol.% D2EHPA + 5 vol.%TBP	RT, pH = 4, O/A = 1, t = 5 min	0.2 M H_2SO_4, O/A = 1	95% Mn, < 2% Li	[79]

(continued)

Table 4 (continued)

Element	Organic phase	Optimized condition	Stripping	Yield (%)	References
Mn(Co/Li)	15 vol.% Co-D2EHPA + 5 vol.% TBP	RT, pH = 3.5, O/A = 1:2, t = 5 min	0.1 M H_2SO_4, O/A = 2	97.1% Mn	[13]
	15% D2EHPA	T = 10 min, T = 25 °C, pH = 3.96–4.02, A/O = 2	0.1 M H_2SO_4, A/O = 1	97% Mn	[23]
Mn(Co/Ni/Li)	25 vol.% Versatic 10 acid+ 20 vol.% PC88A	RT, pH = 4.5, O/A = 1	0.5 M H_2SO_4, O/A = 2	99% Mn, 5.8% Co, 0.8% Ni, 0.3% Li	[119]
	0.4 M D2EHPA + 0.7 M Versatic 10 acid	RT, pH = 4.5, O/A = 1	0.5 M H_2SO_4, O/A = 2	98% Mn, 4.1% Co, 1.1% Ni, 0.3% Li	[120]
Cu/Mn	20 vol.% saponified D2EHPA	pH = 2.6, O/A = 1	–	–	[121]
Cd/Mn	30 vol.% D2EHPA	pH = 2.7,O/A = 2	O/A = 4	77% Cd, 93% Mn	[122]
Mn/Cu/Al	65% saponified D2EHPA + 10% TBP	RT, pH = 4, O/A = 1, D2EHPA/Mn = 4	3 M H_2SO_4, O/A = 1:1.5	90% Mn/Cu/Al	[116]
Ni	10% saponified 1 M Cyanex272	RT, pH = 6.3–6.5, O/A = 1	2 M H_2SO_4, O/A = 1	96% Ni	[68]
Ni(Co)	30% saponified PC-88A	pH = 4.25, O/A = 1	–	80% Ni	[121]
Ni(Mn/Co/Li)	0.23 M LIX 84I + 1.41 M Versatic 10 acid	RT, pH = 5, O/A = 1	3 M H_2SO_4, O/A = 1	98% Ni, 0.3% Co, 0.2% Mn, 0.2% Li	[109]
Cu	10 wt% Acorga M5640	RT, pH = 1.5–1.7, O/A = 1	2 M H_2SO_4, O/A = 1	98.5% Cu	[68]

(continued)

Table 4 (continued)

Element	Organic phase	Optimized condition	Stripping	Yield (%)	References
	10 wt% Acorga M5640	RT, pH = 1.0, O/A = 1, t = 10 min	2 M H_2SO_4, O/A = 1	97% Cu	[62]
	10 vol.% Acorga M5640	RT, pH = 1.5–2.0, O/A = 1, t = 2 h	3 M H_2SO_4, O/A = 1	100% Cu	[110]
	15% LIX 84 IC	pH = 2, O/A = 1	–	99% Cu	[110]
Cu(Co)	L64 + Na_2SO_4 + H_2O (PAN as extracting agent)	pH = 6	0.33 M HNO_3	$\beta_{Cu/Co} = 322$	[123]
Fe/Cu/Al	7% Ionquest 801 + 2% Acorga M5640	T = 40 °C, pH = 4, O/A = 1:2	80 g/L H_2SO_4, O/A = 8	100% Fe/Cu/Al	[115]
Al	10 vol.% PC-88A	RT, pH = 2.5–3.0, O/A = 1, t = 2 h	3 M H_2SO_4, O/A = 1	90% Al	[110]
Zn	30% Cyanex272	pH = 2.5, A/O = 2	1.2 M H_2SO_4 O/A = 2	97% Zn	[122]

the synergistic extraction method can be a candidate to improve the selectivity of Mn over other metals.

Solvent extraction of Cu: It can be discovered from Table 4 that only limited kinds of organic extracting agents are employed during the extraction of Cu^{2+} ions from leaching solutions. And the organic extracting reagents include Acorga M5640 and LIX 84 IC which are chelating extractants. And these extractants always demonstrate an extremely high selectivity over other metal ions. For example, about 100% Cu can be separated from the leaching solution using 10% Acorga M5640 as extractant under optimized experimental conditions of room temperature, pH = 1.5–2.0, O/A = 1 and t = 2 h. It can be found that this extractant works even under an extremely low pH of 1.0, indicating that the extraction of Cu ions from leaching solutions is based on its chelation and other external factors do not work when using chelating extractants for the extraction of Cu. However, it can be also discovered that easy chelating extraction of Cu ions can result in the difficult stripping of Cu from the Cu loaded organic, and about 3 M H_2SO_4 should be used as scrubbing reagent for the further recovery of Cu from the loaded organics. Therefore, we should balance the extraction with stripping process to attain a sound process for the efficient recovery of Cu from leaching solutions.

Removal of impurity ions: It is well known that the leaching solutions may be contaminated by impurity ions, such as Fe, Al and Cu ions. The impurity ions of Fe are usually generated from the metallic shells of spent LIBs during mechanical crushing and grinding processes. The impurity ions of Al are always generated from the peeling off cathode materials from waste cathodes which are coated with Al foils. The Cu ions, however, can be either treated as valuable metals for the purpose of recovery or as impurity ions for the following purification process. As shown in Table 4, 10% PC-88A was adopted as organic extractant for the removal of Al ions and mixed organic extractants of 7% Ionquest 801 and 2% Acorga M5640 were adopted for the selective removal of impurity ions of Al, Fe and Cu ions from leaching solutions. It can be also discovered that about 100% Fe, Cu and Al can be removed under optimized experimental conditions of T = 40 °C, pH = 4 and O/A = 1:2 in the synergistic extraction system of 7% Ionquest 801 and 2% Acorga M5640. In addition, the removal of Al ions from leaching solution, with removal rate of 90%, can be attained under optimized experimental conditions of room temperature, pH = 2.5–3.0, O/A = 1 and t = 2 h. However, the stripping of impurity ions from loaded organics is very difficult for a relatively high concentration of sulfuric acid (e.g. 3 M) will be required for sufficient stripping of impurity ions from the impurity ions loaded organics.

3.3 *Other Separation Methods*

Except the chemical precipitation and solvent extraction processes, there are also some other processes, such as adsorption, ion exchange and membrane separation methods, for the separation and recovery of different metals from leaching solutions. Specific adsorbents are usually used for the selective separation of target metals by adsorption method. The usage of ion exchange resins is also a typical method for the selective recovery of specific valuable metals based on the functional groups in these resins. And functional membranes are usually used for the selective penetration or retention different metals based on size of pore diameter in different membranes. In this chapter, we will focus on the introduction and analysis of the above three separation method used for the separation and recovery of valuable metals from leaching solutions.

Adsorption method: Adsorption is a traditional method for metals separation and recovery in metallurgical and chemical industry. In the recycling of metals values from spent LIBs, the use of this technology is relatively rare, and most of its application is concentrated on the selective extraction of Li from either original or pretreated solutions. One point in this case is the innovative application of low-cost activated carbon (AC) from coconut shell charcoal prepared by chemical and physical activation methods for the selective recovery of Li from complicated solution of Co, Mn, and Ni ions by semicontinuous columns adsorption experiments [41]. It can be concluded from this study that a maximum surface area of 365 m^2/g with total pore volume of 0.148 cm^3/g can be produced by physical activation at 800 °C, and this adsorbent presents a relatively high selectivity towards Li and other heavy metals cannot be co-extracted with Li during the adsorption process. However, it can be also concluded that the adsorption capacity of this adsorbent cannot reach a high level as other typical adsorbents used for metal separation and recovery. Therefore, a balance of adsorption capacity and selectivity towards target valuable metals should be taken into consideration during the adsorption process, and adsorbents with characteristics of high selectivity, high adsorption capacity, easy availability and low price may be welcomed for the preparation of high quality adsorbents.

Ion exchange method: During the separation and extraction of different metals from leaching solution, the functionalized resins are very critical to achieve high selectivity and loaded capacity. Therefore, current studies usually focus on the exploration of resins with specific functions for the extraction and recovery of different metals from the leaching solutions. For example, Guo et al., reported a hydrometallurgical process for the selective recovery of valuable metals (i.e. Cu^{2+}, Co^{2+} and Li^{+}) from leaching solution of spent LIBs using solvent impregnated resins (SIRs) in acidic chloride media [40]. Figure 2 shows the detailed flowsheet for the recovery of valuable metals from spent Li-ion batteries using SIRs. It can be obtained from

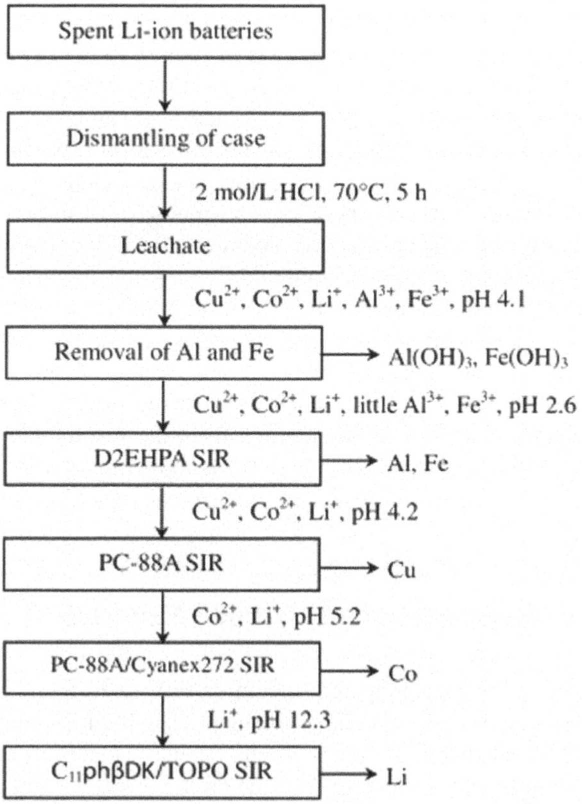

Fig. 2 Flowsheet for the recovery of valuable metals from spent Li-ion batteries using SIRs [40]

this flowsheet that impurity ions of Fe^{3+} and Al^{3+} can be firstly removed from the leachate by precipitation (controlling pH) and column adsorption using SIR containing D2EHPA. Then Cu^{2+} ions can be selectively separated and recovered using a column packed with SIR containing PC-88A at pH = 4.2. Then leaching solution containing Co^{2+} and Li^+ can be obtained after the above separation processes and SIR containing Cyanex 272 (bis(2,4,4-trimethylpentyl) phosphinic acid) was used for the selective recovery of Co^{2+} from the above solution. Finally Li^+ enriched solution can be obtained after the removal or recovery of Al^{3+}, Fe^{3+}, Cu^{2+} and Co^{2+} and SIR containing both 1-phenyl-1,3-tetradecanedione ($C_{11}ph\beta DK$) and tri-n-octylphosphine oxide (TOPO) was applied for the extraction of Li^+. It can be concluded that impurity ions of Al^{3+} and Fe^{3+} ions can be effectively removed and valuable metals of Cu^{2+}, Co^{2+} and Li^+ can be selectively recovered using different functional resins from a relatively complicated solution. Therefore, it is applicable for the adoption of ion exchange method for the separation and recovery of different metals from the leachate containing different metals with sound results.

Membrane separation: As another typical method, the membrane separation usually uses functional membranes for the separation and recovery of valuable metals from leaching solutions under controlled separation conditions. As investigated by Sonoc et al., a Donnan Dialysis based hydrometallurgical process employing cation exchange membrane was proposed for the separation and recovery of Li, Ni, Co and Mn from leaching solution of spent LIBs [124]. In this study, they found that, 94.1% of Li ions can be recovered as high purity lithium carbonate and 99.4% of the transition metal ions (i.e. Ni, Co and Mn) can be recovered as sulphates which can be ready for the following solvent extraction process. Another case in this point is the application of bipolar membrane electrodialysis coupled with chelation for the separation and recovery of valuable metals of Li and Co from spent LIBs [125]. In this study, chelating agent, ethylenediaminetetra- acetic acid (EDTA), was firstly added to a solution containing Li ions and Co ions and almost all Co ions can be chelated with EDTA to form anions, while the lithium ions can hardly be chelated at $pH > 4$. Then the above feed solution was treated using a three-cell-type electrodialysis system, with a unit consisting of two ion-exchange membranes and a bipolar membrane for the transport of Li ions in lithium recovery cell and Co recovery in an applied electric field. It can be obtained from the results that the selectivity for Co and Li in the recovery cell was about 99%, indicating that it is a promising method for the efficient recovery of valuable metals from leaching solutions.

4 Regeneration of Cathode Materials

After above leaching and separation processes, metal enriched leaching solutions or separated products can be obtained. Usually, these mother liquors or raw materials will be then used for preparation of cathode materials by hydrometallurgical method. In this chapter, we mainly focus on the direct re-synthesis of $LiCo_xNi_yMn_zO_2$ (x <

1, $y < 1$, $z < 1$, $x + y + z = 1$) from obtained leaching liquors by co-precipitation, sol-gel and other methods.

4.1 Re-Preparation of Cathode Materials by Co-precipitation Method

The co-precipitation methods usually mix raw materials according to the molar ratio of target cathode materials, and then precipitants are added to the well prepared solution to co-precipitate all required metal ions with expected molar ratio from above leaching solutions to obtain precursor precipitant. Then this precipitant will be subjected for the sequential procedures of purification, drying and calcining to obtain required precursor (i.e. $Co_xNi_yMn_zO$). Finally, the precursor is used for preparation of cathode materials of $LiCo_xNi_yMn_zO_2$. During the co-precipitation process, main parameters affected the quality of products include pH value of mother solution, reaction temperature, stirring speed, raw material concentration etc. Especially, the equilibrium pH of reaction is very important during the precipitation process for the changes of pH value may caused the proportion of the precipitate changes and affected the electrochemical performance of final products. Usually, hydroxide or carbonate compounds (e.g. NaOH, Na_2CO_3) are employed as precipitants for the co-precipitation of metal ions in leaching solutions. During the co-precipitation process, $NH_3 \cdot H_2O$ is usually used as a chelating agent for the complete precipitation of different metal ions.

Recently, co-precipitation method using oxalate, carbonate, and hydroxide were studied to re-prepare the cathode materials of $LiCo_xNi_yMn_zO_2$ with good crystal structure and electrochemical performance. The electric capacity of $LiCo_xNi_yMn_zO_2$ by oxalic acid precipitating method is equivalent with that of hydroxide precipitation. However, the former one present higher capacity retention rate. The controlling of appropriate molar ratio and other experimental conditions is necessary during the preparation of cathode materials of $LiCo_xNi_yMn_zO_2$ from the recycled precursor of $Co_xNi_yMn_zO$. Following presents a detailed hydrometallurgical preparation processes of cathode materials from the leaching solutions by co-preparation method.

An example in this case is the re-synthesis of ternary cathode materials ($LiNi_{1/3}Co_{1/3}Mn_{1/3}O_2$) by hydroxide co-precipitation method as reported by [126]. In this study, the molar ratio of Ni^{2+}, Co^{2+} and Mn^{2+} in leachate is firstly adjusted to 1:1:1 by the addition of sulfates ($NiSO_4$, $CoSO_4$ and $MnSO_4$) and 5 mol/L NaOH and $NH_3 \cdot H_2O$ were used as precipitant and chelating reagent, respectively. Relatively pure precipitates can be obtained under optimized experimental conditions of stirring speed—500 r/min, reaction temperature—60 °C and aging time—24 h. The precipitate was then filtered and dried to obtain $Ni_{1/3}Co_{1/3}Co_{1/3}(OH)_2$ precursor (the detailed chemical reaction equation can be found in Eq. (1)), and Li_2CO_3 was mixed with the precursor under molar ratio of 1.05:1, and then the mixture was calcined in air atmosphere with a temperature of 900 °C and time of 12 h. Finally, the re-synthesized cathode materials of $LiNi_{1/3}Co_{1/3}Mn_{1/3}O_2$ with uniform spherical-like structure can be obtained and were subjected for characterization and electrochemical results.

$$NiSO_4 \cdot 6H_2O + CoSO_4 \cdot 7H_2O + MnSO_4 \cdot H_2O + NH_3 \cdot H_2O + NaOH \rightarrow Ni_xCo_yMn_z(OH)_2 + NH_3 + H_2O \quad (1)$$

The results show that the first discharge specific capacity is 158 mAh/g at 0.1 C and the capacity retention rate after 100 cycles is still higher than 80%. The $Mn(OH)_2$ will be formed during the precipitation process which may be attributed to unstable property of Mn and the oxidation of Mn(II) into MnOOH or MnO_2 without protection of inert atmosphere. This can greatly affect the electrochemical performance of re-synthesized $LiNi_{1/3}Co_{1/3}Mn_{1/3}O_2$ materials [127, 128].

Besides, Jo et al. investigated the effect of residual Li ions on the preparation of precursors $Li_x[Ni_{1/3}Co_{1/3}Mn_{1/3}]_{1-x}(OH)_2$ synthesized by hydroxide co-precipitation method on electrochemical performances [129]. Aqueous solution of $NiSO_4$, $CoSO_4$, $MnSO_4$, and Li_2SO_4 (molar ratio of Ni:Co:Mn = 1:1:1) with a total concentration of 2 mol/L was pumped into a continuously stirred tank reactor under N_2 atmosphere. At the same time, 2 mol/L NaOH and 1 mol/L NH_4OH aqueous solutions were also separately fed into the reactor as precipitant and chelating agent, respectively. The pH of the solution was controlled at 10.5 and the reactor was maintained at a temperature of 50 °C with stirring speed of 1000 rpm for about 70 h. After that, precursor slurry was filtered and washed with distilled water from 0 to 6 times to control the Li content of precursors, and then they were dried in an oven at 80 °C for about 12 h. The dried precursor powders and the stoichiometric amount of Li_2CO_3 were mixed with the precursors thoroughly. The mixed powders were then calcined at 1000 °C for 8 h to synthesize the cathode materials. Three kinds of precursors that contain different amounts of Li were chosen depending on different conditions of the solution composition for the co-precipitation and washing process. They discovered that XRD peaks become more indistinct with an increase of Li amount in precursors, which indicates that the introduction of Li to the precursors reduces the degree of structural perfection. The rate capability of the cathode materials also deteriorates when extra Li exists in the precursors. In summary, electrochemical performances, such as the charge/discharge capacity and the rate capability, is hindered when residual Li is contained in their precursors owing to an increase in the degree of the cation mixing, which was supported by the I(003)/I(104) ratio in the XRD analysis.

Hashem et al. [130] chosen the co-precipitation method to re-synthesize $LiNi_{1/3}Co_{1/3}Mn_{1/3}O_2$ materials and reduce the size of particles with respect to solid-state reaction. Precursor of Ni, Co and Mn oxalate $(Ni_{1/3}Co_{1/3}Mn_{1/3})C_2O_4$, prepared by co-precipitation method using a solution of $Ni(NO_3)_2$, $Co(NO_3)_2$ and $Mn(NO_3)_2$ (cationic molar ratio of Ni:Co:Mn = 1:1:1), can be obtained by dropwise addition of 1 mol/L oxalic acid to the mother liquor for the complete precipitation of oxalate precursor under strong agitation condition. Then the slurry was filtered and dried at 100 °C for 12 h. Stoichiometric amount of lithium hydroxide (LiOH) was mixed thoroughly with the dried powders and calcined at 900 °C for 12 h in air to obtain final product of $LiNi_{1/3}Co_{1/3}Mn_{1/3}O_2$ powders. The surface modification was conducted by mixing 2 g of $LiNi_{1/3}Co_{1/3}Mn_{1/3}O_2$ with 0.1 g sucrose (3 wt%) in an agate mortar for more than 2 h. After an intimate mixing, the powders were heat treated at 600 °C for half-an-hour in air. Similar procedures were also proposed using starch instead of sucrose as carbon source. They also found that the heat treatment time (at 600 °C) is an important parameter since the generated carbon may further react with oxygen in the air for longer exposure duration. A capacity higher than 150 mAh/g can be obtained in voltage range 2.5–4.2 V versus Li^0/Li^+ at current density 1/12–2 C with an increase of 8% for the surface-modified $Li(Ni_{1/3}Co_{1/3}Mn_{1/3})O_2$ cathode material can be also obtained.

A novel process was proposed by the application of oxalic acid leaching method to regenerate waste $Li(Ni_{1/3}Co_{1/3}Mn_{1/3})O_2$ cathodes from spent LIBs [131]. Specifically, 0.6 mol/L oxalic acid was employed as leaching reagent and a fixed amount of waste NCM cathode powders (waste $Li(Ni_{1/3}Co_{1/3}Mn_{1/3})O_2$ cathodes) was placed in a 100 mL three-necked and round-bottomed reactor with an impeller stirrer and a condenser. The whole reactor was equipped in a water bath to maintain reaction temperature at 70 °C with solid to liquid ratio of 20 g/L. After leaching, precipitates with different retention time were filtered and dried. Finally, the precipitates were mixed with a certain amount of Li_2CO_3 and then calcined at 900 °C for 14 h to obtain regenerated cathode materials. For the recovery of lithium in the leaching solution, NaOH was added to adjust pH at 11 to remove residual Ni/Co/Mn ions. Then, leaching solution was condensed to reach a relatively high concentration of Li (>10 g/L), which is essential for the precipitation of Li_2CO_3. Then, saturated Na_2CO_3 was added to the solution under 90 °C. The obtained Li_2CO_3 precipitate was washed with hot water, and then filtrated and dried for 24 h. For comparison, fixed amount of $NiSO_4{\cdot}6H_2O$, $CoSO_4{\cdot}7H_2O$, $MnSO_4{\cdot}H_2O$ (corresponding to molar amount of transition metal contents in spent sample after leaching experiment) and 0.6 mol/L oxalic acid were mixed in the reactor and reacted at 70 °C for 2 h. The detailed flowsheet of the regeneration process is shown in Fig. 3. The regenerated cathode materials after 10 min leaching exhibits the best electrochemical performances, demonstrating the highest initial specific discharge capacity of 168 mAh/g at 0.2 C and 153.7 mAh/g with a high capacity retention of 91.5% after 150 cycles. The excellent electrochemical performances may be attributed to submicrometer particles and voids after calcination, as well as the optimal proportion of elements. This process can make the most of valuable metals in the spent cathodes, with recycling efficiency over 98.5% for Ni, Co, and Mn. Besides, it is also a simple and effective route for the recycling of waste cathodes from spent LIBs.

Carbonate co-precipitation method was used by He et al. [132] for the repreparation of $LiNi_{1/3}Co_{1/3}Mn_{1/3}O_2$ cathode material from waste lithium-ion batteries (LIBs) stream. In their study, obtained waste cathode materials were completely dissolved in a solution of 1 mol/L H_2SO_4 and 1 vol.% H_2O_2. Then molar ratio of Ni^{2+}, Co^{2+} and Mn^{2+} ions was adjusted to 1:1:1 by adding analytically pure sulphates of $NiSO_4{\cdot}6H_2O$, $CoSO_4{\cdot}7H_2O$, and $MnSO_4{\cdot}H_2O$. Then Na_2CO_3 and $NH_3{\cdot}H_2O$ were respectively used as precipitant and chelating reagent for the preparation of precursors, under a controlled pH of 7.5, a stoichiometric ratio of precipitant, a stirring

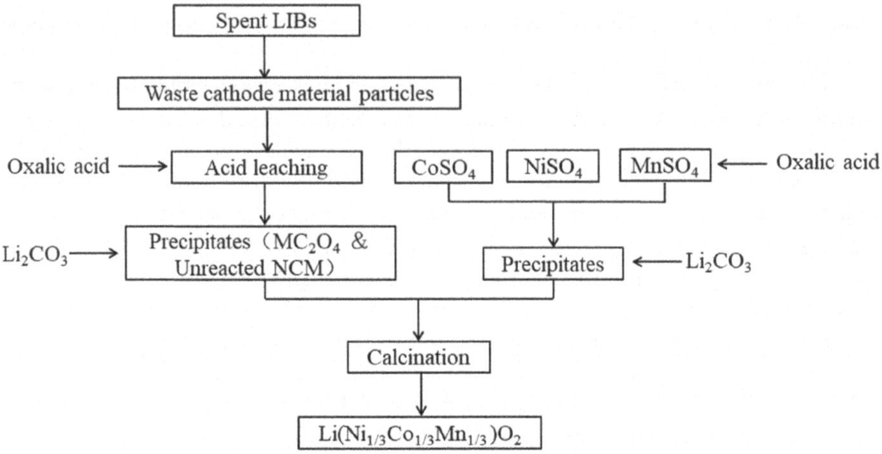

Fig. 3 Flowsheet for the regeneration of $LiNi_{1/3}Co_{1/3}Mn_{1/3}O_2$ from waste cathode materials of spent lithium-ion batteries [131]

speed of 700 rpm, a reaction temperature of 60 °C and a retention time of 12 h. Subsequently, obtained precipitate was thermally treated in an air atmosphere at 500 °C for 5 h to obtain $(Ni_{1/3}Co_{1/3}Mn_{1/3})CO_3$, and then calcined at 900 °C for 12 h after adding appropriate Li salt to obtain $LiNi_{1/3}Co_{1/3}Mn_{1/3}O_2$ cathode materials. Electrochemical performance of regenerated $LiNi_{1/3}Co_{1/3}Mn_{1/3}O_2$ was investigated by continuous charge–discharge cycling and cyclic voltammetry. The results indicate that the regenerated $Ni_{1/3}Co_{1/3}Mn_{1/3}CO_3$ precursor comprises uniform spherical particles with narrower particle-size distribution, which is similar to those of $Ni_{1/3}Co_{1/3}Mn_{1/3}CO_3$ precursors. Moreover, they hold a well-ordered layered structure and low degree of cation mixing. The regenerated $LiNi_{1/3}Co_{1/3}Mn_{1/3}O_2$ shows an initial discharge capacity of 163.5 mAh/g at 0.1 C between 2.7–4.3 V and a discharge capacity of 135.1 mAh/g at 1 C. And the capacity retention ratio is 94.1% after 50 cycles. Even at high rate of 5 C, $LiNi_{1/3}Co_{1/3}Mn_{1/3}O_2$ presents a high capacity of 112.6 mAh/g. These results indicate that electrochemical performance of regenerated $LiNi_{1/3}Co_{1/3}Mn_{1/3}O_2$ is comparable to that of cathode materials synthesized from fresh materials by carbonate co-precipitation.

Moreover, Deng et al. [133] compared different precipitants of Na_2CO_3, $(NH_4)_2CO_3$ and NH_4HCO_3 ($NH_3{\cdot}H_2O$ as chelating reagent) for re-synthesis of $LiNi_{1/3}Co_{1/3}Mn_{1/3}O_2$ materials from spent LIBs. Effects of different precipitants on morphological, structural and electrochemical characteristics of prepared samples were studied. Specific method steps are listed as: (1) An aqueous solution of $NiSO_4$, $CoSO_4$, and $MnSO_4$ (cation molar ratio of Ni:Co:Ni = 1:1:1) was pumped into a continuous stirred tank reactor. At the same time, the solution containing precipitant of Na_2CO_3, $(NH_4)_2CO_3$ or NH_4HCO_3 was also separately added to the reactor; (2) pH of co-precipitation solution was controlled via carefully controlling addition speed of precipitant solutions; (3) The coprecipitation system was maintained at 40 °C with continuous stirring for 12 h; (4) The precipitated powders were filtered and washed, and then they were dried in a vacuum oven at room temperature for 12 h; (5) Obtained precursors were pressed into pellets then calcined at 500 °C for 5 h and were subsequently ground, the obtained powders were mixed with different amounts of LiOH (molar ratio of Li:(Mn + Co + Ni) is 1.10) using a ball mill and pellets were finally remade and then calcinated at 850 °C for 12 h in air. The results show that samples prepared using Na_2CO_3 as precipitant show irregular particle shape and nonuniform particle size, while samples prepared using $(NH_4)_2CO_3$ as precipitant present spherical particle shape and uniform particle size. Among all samples, the one prepared with $(NH_4)_2CO_3$ exhibits the best hexagonal layered structure, which results in its highest discharge capacity and best cycling performance. However, special images of big quasi-spherical union can be obtained for samples prepared using NH_4HCO_3 as precipitant and this may be attributed to hydrolytic side reaction during co-precipitation process, which leads to the less crystallinity and poorer electrochemical performance. Therefore, NH^{4+} plays a key role during the co-precipitation reaction and makes a great impact on the characteristics of $LiNi_{1/3}Co_{1/3}Mn_{1/3}O_2$ cathode materials.

4.2 Re-Preparation of Cathode Materials by Sol-Gel Method

The sol-gel method mainly uses organic acid as chelating reagents and adjusts the molar ratio of metal ions in the leaching solution. At the same time, a sol can be formed by controlling the specific experimental conditions, and then a dry gel can be obtained after evaporation of the above sol. Finally, cathode materials (e.g. $LiNi_{1/3}Co_{1/3}Mn_{1/3}O_2$) can be obtained after calcination of the gel under high temperature.

New cathode materials of $LiCo_{1/3}Ni_{1/3}Mn_{1/3}O_2$ were successfully resynthesized by Li et al. [27] using a sol-gel method from leaching solution of waste cathode materials obtained from spent LIBs (see Fig. 4). The detailed re-preparation procedures are listed as: (1) First, spent LIBs were successively treated with discharging, dismantling, Al removal, drying, calcinating and grinding to obtain scrapped materials of waste cathode materials from spent LIBs. (2) Lactic acid and H_2O_2 were used as leaching reagent and reductant for the complete dissolution of target metals in the leaching solution and this solution will be used as raw materials for the preparation of new cathode materials. (3) Molar ratio of Li^+, Ni^{2+}, Co^{2+} and Mn^{2+} in the leaching solution was adjusted to 1.05:1:1:1; adjust pH to 7 with $NH_3{\cdot}H_2O$ at temperature of 80 °C to obtain sol; stir the sol under moderate agitation time of 5 h and then dry the sol; calcine the sol at a temperature of 450 °C for 5 h to obtain precursor; calcine the precursor with lithium source (e.g. LiOH, Li_2CO_3) at 900 °C for 12 h under air atmosphere to prepare $LiNi_{1/3}Co_{1/3}Mn_{1/3}O_2$. According to the results of electrochemical performance tests, the first discharge specific capacity is 138.2 mAh/g at 0.5 C, with an excellent capacity retention rate of 96% after 100 cycles.

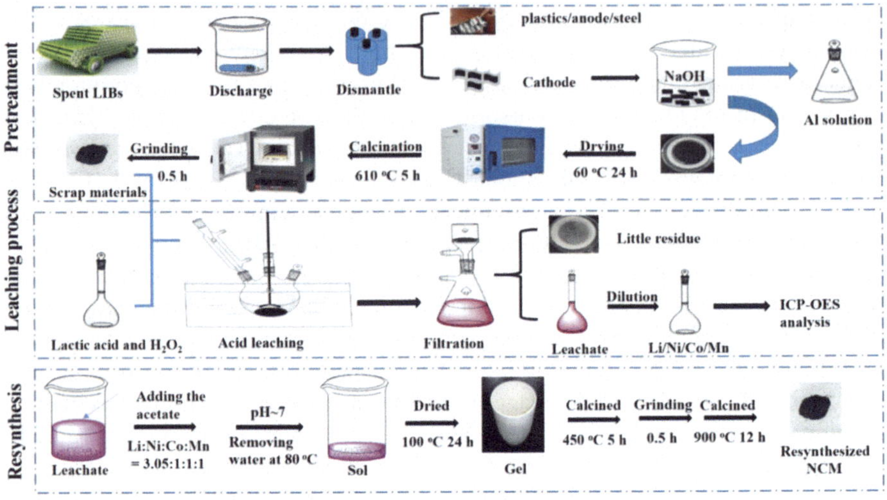

Fig. 4 Flow-sheet for the regeneration of $LiCo_{1/3}Ni_{1/3}Mn_{1/3}O_2$ by sol-gel method. Reprinted with permission from Ref. [27], Copyright @ 2017 American Chemical Society

Besides, citric acid was used as a chelating agent by Yao et al. for the preparation of cathode materials using nitrates of Li, Co, Ni, and Mn as raw materials leached from spent LIBs [134]. The molar ratio of Li^+, Ni^{2+}, Co^{2+}, and Mn^{2+} is adjusted to 3.05:1:1:1, under a controlled pH of 8 using $NH_3{\cdot}H_2O$ as acidity regulator to obtain sol. Then the sol was heated at a temperature of 80 °C to obtain precursor, and the precursor was calcine at 350 °C for 2 h and then at a temperature of 750 °C for 2 h to in air atmosphere. The ternary cathode materials can be prepared by secondary calcination at 750 °C for 24 h. The results of electrochemical performance tests show that the first discharge specific capacity was 147 m Ah/g at 1 C rate, and the capacity retention rate after 100 cycles was 93%. Finally, it is found that citric acid can provide more carboxyl groups as a chelating reagent than that of tartaric acid, which can promote the sol-gel preparation process of cathode materials with smaller particle size and improve electrochemical performances of cathode materials [135].

It is also reported by Xia et al. [136] that citric acid and vinyl alcohol can be also used as chelating reagents for re-synthesis of $LiNi_{1/3}Co_{1/3}Mn_{1/3}O_2$ from nitrates leaching solution containing $LiNO_3$, $Mn(NO_3)_2$ (50 wt% of $Mn(NO_3)_2$), $Co(NO_3)_2{\cdot}6H_2O$ and $Ni(NO_3)_2{\cdot}6H_2O$. Firstly, the sol can be obtained at a temperature of 90 °C for 1 h under moderate agitation. The sol was then dried and calcined at 900 °C for 12 h to prepare $LiNi_{1/3}Co_{1/3}Mn_{1/3}O_2$. Characterization results indicate that the obtained cathode materials present a small particle size and good rate performances. It can be also discovered that the preparation of cathode materials of $LiNi_{1/3}Co_{1/3}Mn_{1/3}O_2$ is a relatively simple recycling process, with advantages of simplified process equipment, easy industrialization, and controllable purity of the precursor. However, this method is also insufficient for the removal of impurities under harsh precipitation conditions. The resynthesize of $LiNi_{1/3}Co_{1/3}Mn_{1/3}O_2$

cathode materials by sol-gel method has advantages of fine fraction (e.g. smaller submicron or nanometer structure) and uniform mixing and high purity, as well as disadvantages of complicated operation, high cost, reproducibility and not easy to industrialize, etc. [137].

4.3 Direct Regeneration of Cathode Materials

The above two recycling methods are involved with the dissolving of waste cathode materials into different leaching solutions and then re-preparation of new cathode materials from the recycled leaching solutions, which may consume a large amount of chemicals (e.g. acid, chelating reagents) and the tedious processes may also discourage the efficient recycling of different metals from spent LIBs. Therefore, several studies have attempted the direct restoration and recycling of waste cathode materials from spent LIBs. In this section, authors will concentrate on the introduction and discussion of several existed techniques to explore current direct recycling processes of cathode materials from spent LIBs [138–143].

A case in this point is the re-preparation of new cathode materials $LiNi_{1/3}Co_{1/3}Mn_{1/3}O_2–V_2O_5$ by simultaneous recycling of waste cathode materials in spent LIBs and vanadium-bearing slag as reported by Meng et al. [140]. Spent NCM (waste cathode materials of $LiNi_{1/3}Co_{1/3}Mn_{1/3}O_2$) and vanadium-bearing slag were used as raw materials for the sustainable preparation of new cathode materials of $LiNi_{1/3}Co_{1/3}Mn_{1/3}O_2–V_2O_5$ (see Fig. 5). According to the results of characterization and electrochemical performance tests, it can be concluded that a kind of new cathode materials (i.e. $LiNi_{1/3}Co_{1/3}Mn_{1/3}O_2–V_2O_5$) was successfully prepared with a specific

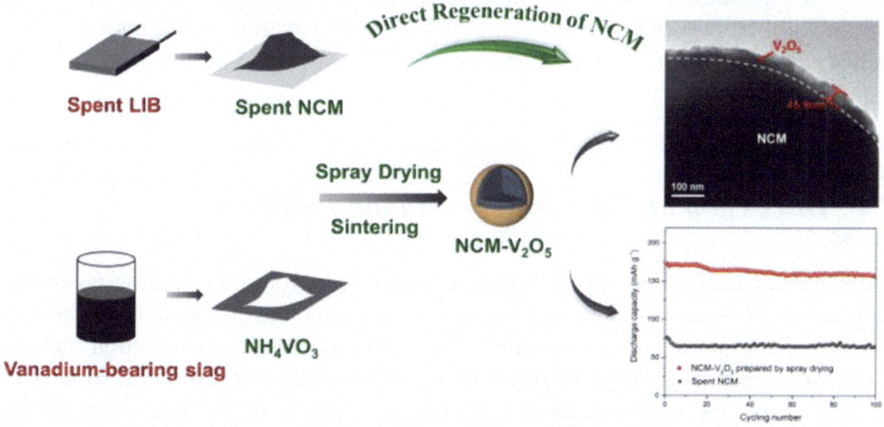

Fig. 5 Re-preparation of cathode materials $LiNi_{1/3}Co_{1/3}Mn_{1/3}O_2–V_2O_5$ by simultaneous recycling of waste cathode materials in spent LIBs and vanadium-bearing slag. Reprinted with permission from Ref. [140], Copyright @ 2017 American Chemical Society

capacity of 156.3 mAh·g^{-1} and a capacity retention of 90.6% after 100 cycles at 0.1 C. Besides, it can be also discovered that spray drying method is an effective method during the coating of V_2O_5 on the thin layer of NCM ($LiNi_{1/3}Co_{1/3}Mn_{1/3}O_2$) surface. This direct regeneration process of cathode materials can be a short-cut process with advantages of lower energy consumption, lower environmental impacts and high-performance cathode materials than traditional metallurgical methods, which may be capable for the practical production of new batteries.

Moreover, a green recycling process for the direct regeneration of waste $LiCoO_2$ from spent LIBs was proposed by Nie et al., by solid state synthesis method using Li_2CO_3 as additive [141]. Both characterization (XRD analysis) and electrochemical performances results indicate that regenerated $LiCoO_2$ powders are similar with those of commercial $LiCoO_2$, when waste LiCoO2 was thermally treated at 900 °C. It can be also concluded from the results of electrochemical performances that the regenerated cathode materials hold a discharge capacity of 152.4 mAh·g^{-1}, an attenuation rate of capacity during every cycle of 0.0313 mAh·g^{-1}, a plateau retention at 3.6 V of 99.08% and attenuation rate of plateau during every cycle of 0.0463 mAh·g^{-1} under the optimized thermal treatment condition of 900 °C. For the recycling of waste $LiFePO_4$, a kind of widely used cathode material in EV/HEV, Song et al. established a direct restoration technology for the direct recycling of spent $LiFePO_4$ batteries using solid phase sintering method [142]. The waste LIBs were firstly subjected for the pretreatment procedures, including separating the cathodes and anodes by manual dismantling and peeling off Al foils by soaking the cathodes in organic solvent of DMAC for 30 min at 30 °C. Then the peeled off waste cathode materials were treated by doping new LiFePO4 at different ratios at suitable temperatures. The results obtained from electrochemical performances indicate that the discharge capacity of regenerated $LiFePO_4$ can reach over 120 mAh·g^{-1} at 0.1 C, with a highest discharge capacity of 144 mAh·g^{-1} under doping ratio of 3:7 and calcination temp. of 700 °C. Besides, it can be also discovered that lower temperatures (i.e. 600 and 700 °C) will be beneficial for the rate capabilities and cycling performances of the cathode materials made from regenerated $LiFePO_4$, which can meet the primary requirements of battery reuse.

4.4 Discussion of Current Regeneration Processes

It can be concluded from the above discussion that current regeneration processes include the co-precipitation method, sol-gel method and direct regeneration method for re-preparation of cathode materials. Among them, the co-precipitation method is the most sophisticated technology with the most concerns and studies, which can be attributed to its credible cathode materials products from the wastes, capable for different kinds of waste stream and easy to control the detailed recycling conditions etc. However, it can be also discovered that this process is also confronted with several disadvantages, such as the large consumption of chemicals, enhanced cost by the tedious recycling process, secondary contamination caused during the recovery process etc. For the sol-gel method, however, it can greatly alleviate these disadvantages involved with the co-precipitation method in terms of secondary pollution, large chemical consumption and tedious recycling process. For example, organic acids are usually employed during the regeneration of cathode materials, which can greatly reduce adverse impact on eco-system when using inorganic acids. Besides, mild reaction conditions and shorted processes adopted by the sol-gel method can also cut down the cost of equipment and chemical reagents. The direct regeneration method can be wise choice for the renovation or regeneration of the waste cathode materials with relatively simple components. Furthermore, this method can also shorten the recycling processes and reduce the usage of chemical reagents to the maximum extent. However, this regeneration method also comes across a severe drawback that this regeneration method cannot be applicable to the more complicated waste streams, such as $LiNi_xCo_yMn_zO_2$, a mixture of several kinds of waste cathode materials etc., which may discourage its wide application and industrialization. Therefore, the recycling processes with wider applicability, away from secondary pollution, short-cut process and mild reaction condition may be the future trends for the re-preparation of cathode materials from spent LIBs.

5 Conclusions

For the purpose of recycling of different metals from spent LIBs, relatively comprehensive introduction and analysis of the current hydrometallurgical processes was presented in this chapter. Based on above presentation, several conclusions can be obtained concerning hydrometallurgical processes involved with reductive leaching, metal separation and cathode materials regeneration. Firstly, the waste cathode materials obtained after the pretreatment procedure of spent LIBs can be effectively dissolved or recovered after leaching process. Both mineral acids and organic acids can be used as leaching reagents during the leaching reactions, and it is suggested mineral acids show higher leaching ability towards different metals while organic acids illustrate higher selectivity to valuable metals, indicating that we should balance the leaching ability and selectivity to choose the most suitable acid during the leaching of target metals. It can be also discovered from analysis of bioleaching that the direct bioleaching process may illustrate higher leaching efficiencies than that of indirect bioleaching process, and there are still definite room for improvements in terms of leaching efficiency, selectivity and applicability for the bioleaching processes. For the separation and recovery processes, we should adopt the appropriate separation methods (such as chemical precipitation, solvent extraction) for the recovery of each metal from the leaching solutions, and it is suggested that a combination of two or more separation methods should be also taken into consideration for separation and recovery of different metals from leaching solution constituted by more complicated metal ions. Finally, the regeneration of cathode materials from spent LIBs is an important part for the recycling processes, and it can be concluded that co-precipitation method, sol-gel method and direct regeneration method are effective ways for the re-preparation of cathode materials from spent LIBs. And it is suggested that short-cut process, low reagent consumption, mild preparation conditions and enhanced electrochemical performances may be major objectives for the regeneration processes. It is expected that a well-round introduction and discussion can be established concerning the hydrometallurgical processes for different metals recycling from spent LIBs in aspects of leaching, metal separation/recovery and cathode materials re-preparation in this chapter.

References

1. J. Dewulf, G. Van der Vorst, K. Denturck, H. Van Langenhove, W. Ghyoot, J. Tytgat, K. Vandeputte, Recycling rechargeable lithium ion batteries: Critical analysis of natural resource savings. Resour. Conserv. Recycl. **54**, 229–234 (2010)
2. T. Zhang, Y. He, F. Wang, L. Ge, X. Zhu, H. Li, Chemical and process mineralogical characterizations of spent lithium-ion batteries: an approach by multi-analytical techniques. Waste Manage. **34**, 1051–1058 (2014)
3. J. Li, G. Wang, Z. Xu, Generation and detection of metal ions and volatile organic compounds (VOCs) emissions from the pretreatment processes for recycling spent lithium-ion batteries. Waste Manage. **52**, 221–227 (2016)
4. Y. Yu, B. Chen, K. Huang, X. Wang, D. Wang, Environmental impact assessment and end-of-life treatment policy analysis for Li-ion batteries and Ni-MH batteries. Int. J. Environ. Res. Publ. Health **11**, 3185–3198 (2014)
5. X. Zeng, J. Li, L. Liu, Solving spent lithium-ion battery problems in China: opportunities and challenges. Renew. Sustain. Energy Rev. **52**, 1759–1767 (2015)
6. X. Wang, G. Gaustad, C.W. Babbitt, C. Bailey, M.J. Ganter, B.J. Landi, Economic and environmental characterization of an evolving Li-ion battery waste stream. J. Environ. Manage. **135**, 126–134 (2014)
7. A.K. Awasthi, J. Li, An overview of the potential of eco-friendly hybrid strategy for metal recycling from WEEE. Resour. Conserv. Recycl. **126**, 228–239 (2017)
8. K. Richa, C.W. Babbitt, G. Gaustad, X. Wang, A future perspective on lithium-ion battery waste flows from electric vehicles. Resour. Conserv. Recycl. **83**, 63–76 (2014)
9. L. Gaines, The future of automotive lithium-ion battery recycling: Charting a sustainable course. Sustain. Mat. Technol. **1–2**, 2–7 (2014)
10. X. Wang, G. Gaustad, C.W. Babbitt, Targeting high value metals in lithium-ion battery recycling via shredding and size-based separation. Waste Manage. **51**, 204–213 (2016)
11. Y. Yang, X. Zheng, H. Cao, C. Zhao, X. Lin, P. Ning, Y. Zhang, W. Jin, Z. Sun, A closed-loop process for selective metal recovery from spent lithium iron phosphate batteries through mechanochemical activation. ACS Sustain. Chem. Eng. (2017)
12. G.-X. Ren, S.-W. Xiao, M.-Q. Xie, B. Pan, J. Chen, F.-G. Wang, X. Xia, Recovery of valuable metals from spent lithium ion batteries by smelting reduction process based on FeO–SiO 2–Al 2 O 3 slag system. Trans. Nonferrous Met. Soc. China **27**, 450–456 (2017)
13. X. Chen, Y. Chen, T. Zhou, D. Liu, H. Hu, S. Fan, Hydrometallurgical recovery of metal values from sulfuric acid leaching liquor of spent lithium-ion batteries. Waste Manage. **38**, 349–356 (2015)
14. Y. Xin, X. Guo, S. Chen, J. Wang, F. Wu, B. Xin, Bioleaching of valuable metals Li Co, Ni and Mn from spent electric vehicle Li-ion batteries for the purpose of recovery. J. Clean. Prod. **116**, 249–258 (2016)

15. J. Li, G. Wang, Z. Xu, Environmentally-friendly oxygen-free roasting/wet magnetic separation technology for in situ recycling cobalt, lithium carbonate and graphite from spent LiCoO2/graphite lithium batteries. J. Hazard. Mat. **302**, 97–104 (2016)
16. M.-M. Wang, C.-C. Zhang, F.-S. Zhang, An environmental benign process for cobalt and lithium recovery from spent lithium-ion batteries by mechanochemical approach. Waste Manage. **51**, 239–244 (2016)
17. J. Guan, Y. Li, Y. Guo, R. Su, G. Gao, H. Song, H. Yuan, B. Liang, Z. Guo, Mechanochemical process enhanced cobalt and lithium recycling from wasted lithium-ion batteries. ACS Sustain. Chem. Eng. **5**, 1026–1032 (2016)
18. L. Sun, K. Qiu, Vacuum pyrolysis and hydrometallurgical process for the recovery of valuable metals from spent lithium-ion batteries. J. Hazard. Mater. **194**, 378–384 (2011)
19. Y. Guo, F. Li, H. Zhu, G. Li, J. Huang, W. He, Leaching lithium from the anode electrode materials of spent lithium-ion batteries by hydrochloric acid (HCl). Waste Manage. **51**, 227–233 (2016)
20. Z. Niu, Y. Zou, B. Xin, S. Chen, C. Liu, Y. Li, Process controls for improving bioleaching performance of both Li and Co from spent lithium ion batteries at high pulp density and its thermodynamics and kinetics exploration. Chemosphere **109**, 92–98 (2014)
21. X. Chen, C. Luo, J. Zhang, J. Kong, T. Zhou, Sustainable recovery of metals from spent lithium-ion batteries: a green process. ACS Sustain. Chem. Eng. **3**, 3104–3113 (2015)
22. X. Chen, B. Xu, T. Zhou, D. Liu, H. Hu, S. Fan, Separation and recovery of metal values from leaching liquor of mixed-type of spent lithium-ion batteries. Sep. Purif. Technol. **144**, 197–205 (2015)
23. X. Chen, T. Zhou, J. Kong, H. Fang, Y. Chen, Separation and recovery of metal values from leach liquor of waste lithium nickel cobalt manganese oxide based cathodes. Sep. Purif. Technol. **141**, 76–83 (2015)
24. J. Xiao, J. Li, Z. Xu, Recycling metals from lithium ion battery by mechanical separation and vacuum metallurgy. J. Hazard. Mater. **338**, 124–131 (2017)
25. A. Chagnes, B. Pospiech, A brief review on hydrometallurgical technologies for recycling spent lithium-ion batteries. J. Chem. Technol. Biotechnol. **88**, 1191–1199 (2013)
26. L.P. He, S.Y. Sun, Y.Y. Mu, X.F. Song, J.G. Yu, Recovery of lithium nickel, cobalt, and manganese from spent lithium-ion batteries using l-tartaric acid as a leachant. Acs Sustain. Chem. Eng. **5**, 714–721 (2017)
27. L. Li, E. Fan, Y. Guan, X. Zhang, Q. Xue, L. Wei, F. Wu, R. Chen, Sustainable recovery of cathode materials from spent lithium-ion batteries using lactic acid leaching system. ACS Sustain. Chem. Eng. **5**, 5224–5233 (2017)
28. R.-C. Wang, Y.-C. Lin, S.-H. Wu, A novel recovery process of metal values from the cathode active materials of the lithium-ion secondary batteries. Hydrometallurgy **99**, 194–201 (2009)
29. S.-G. Zhu, W.-Z. He, G.-M. Li, X. Zhou, X.-J. Zhang, J.-W. Huang, Recovery of Co and Li from spent lithium-ion batteries by combination method of acid leaching and chemical precipitation. Trans. Nonferrous Met. Soc. China **22**, 2274–2281 (2012)
30. L. Chen, X. Tang, Y. Zhang, L. Li, Z. Zeng, Y. Zhang, Process for the recovery of cobalt oxalate from spent lithium-ion batteries. Hydrometallurgy **108**, 80–86 (2011)
31. L. Li, J. Ge, F. Wu, R. Chen, S. Chen, B. Wu, Recovery of cobalt and lithium from spent lithium ion batteries using organic citric acid as leachant. J. Hazard. Mater. **176**, 288–293 (2010)
32. W. Lv, Z. Wang, H. Cao, Y. Sun, Y. Zhang, Z. Sun, A critical review and analysis on the recycling of spent lithium-ion batteries. ACS Sustain. Chem. Eng. **6**, 1504–1521 (2018)
33. W. Zhang, C. Xu, W. He, G. Li, J. Huang, A review on management of spent lithium ion batteries and strategy for resource recycling of all components from them. Waste Manage. Res. **36**, 99–112 (2018)
34. J. Xu, H.R. Thomas, R.W. Francis, K.R. Lum, J. Wang, B. Liang, A review of processes and technologies for the recycling of lithium-ion secondary batteries. J. Power Sour. **177**, 512–527 (2008)

35. X. Zeng, J. Li, N. Singh, Recycling of spent lithium-ion battery: a critical review. Crit. Rev. Environ. Sci. Technol. **44**, 1129–1165 (2014)
36. J. Ordonez, E.J. Gago, A. Girard, Processes and technologies for the recycling and recovery of spent lithium-ion batteries. Renew. Sustain. Energy Rev. **60**, 195–205 (2016)
37. L. Ma, Z. Nie, X. Xi, X. Li, Theoretical simulation and experimental study on nickel, cobalt, manganese separation in complexation–precipitation system. Sep. Purif. Technol. **108**, 124–132 (2013)
38. F. Wang, F. He, J. Zhao, N. Sui, L. Xu, H. Liu, Extraction and separation of cobalt(II), copper(II) and manganese(II) by Cyanex272, PC-88A and their mixtures. Sep. Purif. Technol. **93**, 8–14 (2012)
39. L. Ma, X. Xi, K. Wang, L. Zhao, Adsorption of Li by a lithium ion-sieve using a buffer system and application for the recovery of Li from a spent lithium-ion battery, 1–19 (2018)
40. F. Guo, S. Nishihama, K. Yoshizuka, Selective recovery of valuable metals from spent Li-ion batteries using solvent-impregnated resins. Environ. Technol. **34**, 1307–1317 (2013)
41. C.W. Purnomo, E.P. Kesuma, I. Perdana, M. Aziz, Lithium recovery from spent Li-ion batteries using coconut shell activated carbon. Waste Manage. **79**, 454–461 (2018)
42. X. Chen, L. Cao, D. Kang, J. Li, T. Zhou, H. Ma, Recovery of valuable metals from mixed types of spent lithium ion batteries. Part II: Selective extraction of lithium. Waste Manage. **80**, 198–210 (2018)
43. X. Chen, H. Ma, C. Luo, T. Zhou, Recovery of valuable metals from waste cathode materials of spent lithium-ion batteries using mild phosphoric acid. J. Hazard. Mater. **326**, 77–86 (2017)
44. Y. Yao, M. Zhu, Z. Zhao, B. Tong, Y. Fan, Z. Hua, Hydrometallurgical processes for recycling spent lithium-ion batteries: a critical review. ACS Sustain. Chem. Eng. (2018)
45. A.A. Nayl, R.A. Elkhashab, S.M. Badawy, M.A. El-Khateeb, Acid leaching of mixed spent Li-ion batteries. Arab. J. Chem. **10**, S3632–S3639 (2017)
46. H. Li, S.Z. Xing, Y. Liu, F.J. Li, H. Guo, G. Kuang, Recovery of lithium, iron, and phosphorus from spent LiFePO4 batteries using stoichiometric sulfuric acid leaching system. Acs Sustain. Chem. Eng. **5**, 8017–8024 (2017)
47. P. Zhang, T. Yokoyama, O. Itabashi, Y. Wakui, T.M. Suzuki, K. Inoue, Hydrometallurgical process for recovery of metal values from spent nickel-metal hydride secondary batteries. Hydrometallurgy **50**, 61–75 (1998)
48. L. Yao, H. Yao, G. Xi, Y. Feng, Recycling and synthesis of LiNi1/3Co1/3Mn1/3O2 from waste lithium ion batteries using d, l-malic acid. RSC Adv. **6**, 17947–17954 (2016)
49. G.P. Nayaka, K.V. Pai, G. Santhosh, J. Manjanna, Dissolution of cathode active material of spent Li-ion batteries using tartaric acid and ascorbic acid mixture to recover Co. Hydrometallurgy **161**, 54–57 (2016)
50. L. Li, W. Qu, X. Zhang, J. Lu, R. Chen, F. Wu, K. Amine, Succinic acid-based leaching system: A sustainable process for recovery of valuable metals from spent Li-ion batteries. J. Power Sour. **282**, 544–551 (2015)
51. E.G. Pinna, M.C. Ruiz, M.W. Ojeda, M.H. Rodriguez, Cathodes of spent Li-ion batteries: Dissolution with phosphoric acid and recovery of lithium and cobalt from leach liquors. Hydrometallurgy **167**, 66–71 (2017)
52. Z.X. Liu, Z.L. Yin, S.F. Xiong, Y.G. Chen, Q.Y. Chen, Leaching and kinetic modeling of calcareous bornite in ammonia ammonium sulfate solution with sodium persulfate. Hydrometallurgy **144–145**, 86–90 (2014)
53. F. Pagnanelli, E. Moscardini, G. Granata, S. Cerbelli, L. Agosta, A. Fieramosca, L. Toro, Acid reducing leaching of cathodic powder from spent lithium ion batteries: Glucose oxidative pathways and particle area evolution. J. Ind. Eng. Chem. **20**, 3201–3207 (2014)
54. X. Chen, C. Guo, H. Ma, J. Li, T. Zhou, L. Cao, D. Kang, Organic reductants based leaching: a sustainable process for the recovery of valuable metals from spent lithium ion batteries. Waste Manage. **75**, 459–468 (2018)
55. K. Tanong, L. Coudert, G. Mercier, J.-F. Blais, Recovery of metals from a mixture of various spent batteries by a hydrometallurgical process. J. Environ. Manage. **181**, 95–107 (2016)

56. S.P. Barik, G. Prabaharan, L. Kumar, Leaching and separation of Co and Mn from electrode materials of spent lithium-ion batteries using hydrochloric acid: Laboratory and pilot scale study. J. Clean. Prod. **147**, 37–43 (2017)
57. Z. Takacova, T. Havlik, F. Kukurugya, D. Orac, Cobalt and lithium recovery from active mass of spent Li-ion batteries: theoretical and experimental approach. Hydrometallurgy **163**, 9–17 (2016)
58. J. Li, P. Shi, Z. Wang, C. Yao, C.C. Chang, A combined recovery process of metals in spent lithium-ion batteries. Chemosphere **77**, 1132–1136 (2009)
59. M. Joulié, R. Laucournet, E. Billy, Hydrometallurgical process for the recovery of high value metals from spent lithium nickel cobalt aluminum oxide based lithium-ion batteries. J. Power Sour. **247**, 551–555 (2014)
60. P. Meshram, B.D. Pandey, T.R. Mankhand, Recovery of valuable metals from cathodic active material of spent lithium ion batteries: Leaching and kinetic aspects. Waste Manage. **45**, 306–313 (2015)
61. R. Zheng, L. Zhao, W. Wang, Y. Liu, Q. Ma, D. Mu, R. Li, C. Dai, Optimized Li and Fe recovery from spent lithium-ion batteries via a solution-precipitation method. RSC Adv. **6**, 43613–43625 (2016)
62. J. Nan, D. Han, X. Zuo, Recovery of metal values from spent lithium-ion batteries with chemical deposition and solvent extraction. J. Power Sour. **152**, 278–284 (2005)
63. M.K. Jha, A. Kumari, A.K. Jha, V. Kumar, J. Hait, B.D. Pandey, Recovery of lithium and cobalt from waste lithium ion batteries of mobile phone. Waste Manage. **33**, 1890–1897 (2013)
64. B. Swain, J. Jeong, J.C. Lee, G.H. Lee, J.S. Sohn, Hydrometallurgical process for recovery of cobalt from waste cathodic active material generated during manufacturing of lithium ion batteries. J. Power Sour. **167**, 536–544 (2007)
65. J. Kang, G. Senanayake, J. Sohn, S.M. Shin, Recovery of cobalt sulfate from spent lithium ion batteries by reductive leaching and solvent extraction with Cyanex 272. Hydrometallurgy **100**, 168–171 (2010)
66. D.A. Ferreira, L.M.Z. Prados, D. Majuste, M.B. Mansur, Hydrometallurgical separation of aluminium, cobalt, copper and lithium from spent Li-ion batteries. J. Power Sour. **187**, 238–246 (2009)
67. G. Dorella, M.B. Mansur, A study of the separation of cobalt from spent Li-ion battery residues. J. Power Sour. **170**, 210–215 (2007)
68. J. Nan, D. Han, M. Yang, M. Cui, X. Hou, Recovery of metal values from a mixture of spent lithium-ion batteries and nickel-metal hydride batteries. Hydrometallurgy **84**, 75–80 (2006)
69. L.-P. He, S.-Y. Sun, X.-F. Song, J.-G. Yu, Leaching process for recovering valuable metals from the LiNi1/3Co1/3Mn1/3O2 cathode of lithium-ion batteries. Waste Manage. **64**, 171–181 (2017)
70. S. Kim, D. Yang, K. Rhee, J. Sohn, Recycling process of spent battery modules in used hybrid electric vehicles using physical/chemical treatments. Res. Chem. Intermediat **40**, 2447–2456 (2014)
71. C.K. Lee, K.-I. Rhee, Reductive leaching of cathodic active materials from lithium ion battery wastes. Hydrometallurgy **68**, 5–10 (2003)
72. L. Li, R. Chen, F. Sun, F. Wu, J. Liu, Preparation of LiCoO2 films from spent lithium-ion batteries by a combined recycling process. Hydrometallurgy **108**, 220–225 (2011)
73. Q. Meng, Y. Zhang, P. Dong, Use of glucose as reductant to recover Co from spent lithium ions batteries. Waste Manage. **64**, 214–218 (2017)
74. G.P. Nayaka, J. Manjanna, K.V. Pai, R. Vadavi, S.J. Keny, V.S. Tripathi, Recovery of valuable metal ions from the spent lithium-ion battery using aqueous mixture of mild organic acids as alternative to mineral acids. Hydrometallurgy **151**, 73–77 (2015)
75. L. Li, J. Lu, Y. Ren, X.X. Zhang, R.J. Chen, F. Wu, K. Amine, Ascorbic-acid-assisted recovery of cobalt and lithium from spent Li-ion batteries. J. Power Sour. **218**, 21–27 (2012)
76. L. Sun, K. Qiu, Organic oxalate as leachant and precipitant for the recovery of valuable metals from spent lithium-ion batteries. Waste Manage. **32**, 1575–1582 (2012)

77. X. Zeng, J. Li, B. Shen, Novel approach to recover cobalt and lithium from spent lithium-ion battery using oxalic acid. J. Hazard. Mater. **295**, 112–118 (2015)
78. X. Chen, D. Kang, L. Cao, J. Li, T. Zhou, H. Ma, Separation and recovery of valuable metals from spent lithium ion batteries: Simultaneous recovery of Li and Co in a single step. Sep. Purif. Technol. **210**, 690–697 (2019)
79. X. Chen, T. Zhou, Hydrometallurgical process for the recovery of metal values from spent lithium-ion batteries in citric acid media. Waste Manage. Res. **32**, 1083–1093 (2014)
80. W. Gao, J. Song, H. Cao, X. Lin, X. Zhang, X. Zheng, Y. Zhang, Z. Sun, Selective recovery of valuable metals from spent lithium-ion batteries—Process development and kinetics evaluation. J. Clean. Prod. **178**, 833–845 (2018)
81. S. Natarajan, A.B. Boricha, H.C. Bajaj, Recovery of value-added products from cathode and anode material of spent lithium-ion batteries. Waste Manage. **77**, 455–465 (2018)
82. Y. Yang, X. Meng, H. Cao, X. Lin, C. Liu, Y. Sun, Y. Zhang, Z. Sun, Selective recovery of lithium from spent lithium iron phosphate batteries: a sustainable process. Green Chem. **20**, 3121–3133 (2018)
83. L. Li, J. Ge, R. Chen, F. Wu, S. Chen, X. Zhang, Environmental friendly leaching reagent for cobalt and lithium recovery from spent lithium-ion batteries. Waste Manage. **30**, 2615–2621 (2010)
84. C. Sun, L. Xu, X. Chen, T. Qiu, T. Zhou, Sustainable recovery of valuable metals from spent lithium-ion batteries using DL-malic acid: Leaching and kinetics aspect. Waste Manage. Res. **36**, 113–120 (2018)
85. L. Li, Y. Bian, X. Zhang, Q. Xue, E. Fan, F. Wu, R. Chen, Economical recycling process for spent lithium-ion batteries and macro- and micro-scale mechanistic study. J. Power Sour. **377**, 70–79 (2018)
86. W.F. Gao, X.H. Zhang, X.H. Zheng, X. Lin, H.B. Cao, Y. Zhi, Z. Sun, Lithium carbonate recovery from cathode scrap of spent lithium-ion battery: a closed-loop process. Environ. Sci. Technol. **51**, 1662–1669 (2017)
87. X. Zhang, H. Cao, Y. Xie, P. Ning, H. An, H. You, F. Nawaz, A closed-loop process for recycling LiNi1/3Co1/3Mn1/3O2 from the cathode scraps of lithium-ion batteries: Process optimization and kinetics analysis. Sep. Purif. Technol. **150**, 186–195 (2015)
88. G.P. Nayaka, K.V. Pai, G. Santhosh, J. Manjanna, Recovery of cobalt as cobalt oxalate from spent lithium ion batteries by using glycine as leaching agent. J. Environ. Chem. Eng. **4**, 2378–2383 (2016)
89. G.P. Nayaka, K.V. Pai, J. Manjanna, S.J. Keny, Use of mild organic acid reagents to recover the Co and Li from spent Li-ion batteries. Waste Manage. **51**, 234–238 (2016)
90. L. Li, J.B. Dunn, X.X. Zhang, L. Gaines, R.J. Chen, F. Wu, K. Amine, Recovery of metals from spent lithium-ion batteries with organic acids as leaching reagents and environmental assessment. J. Power Sour. **233**, 180–189 (2013)
91. B.K. Biswal, U.U. Jadhav, M. Madhaiyan, L. Ji, E.-H. Yang, B. Cao, Biological leaching and chemical precipitation methods for recovery of Co and Li from spent lithium-ion batteries. ACS Sustain. Chem. Eng. **6**, 12343–12352 (2018)
92. N.J. Boxall, K.Y. Cheng, W. Bruckard, A.H. Kaksonen, Application of indirect non-contact bioleaching for extracting metals from waste lithium-ion batteries. J. Hazard. Mater. **360**, 504–511 (2018)
93. J. Wang, B. Tian, Y. Bao, C. Qian, Y. Yang, T. Niu, B. Xin, Functional exploration of extracellular polymeric substances (EPS) in the bioleaching of obsolete electric vehicle LiNixCoyMn1-x-yO2 Li-ion batteries. J. Hazard. Mater. **354**, 250–257 (2018)
94. Y. Zheng, W. Song, W.-T. Mo, L. Zhou, J.-W. Liu, Lithium fluoride recovery from cathode material of spent lithium-ion battery. RSC Adv. **8**, 8990–8998 (2018)
95. Y. Song, Z. Zhao, Recovery of lithium from spent lithium-ion batteries using precipitation and electrodialysis techniques. Sep. Purif. Technol. **206**, 335–342 (2018)
96. G. Granata, E. Moscardini, F. Pagnanelli, F. Trabucco, L. Toro, Product recovery from Li-ion battery wastes coming from an industrial pre-treatment plant: Lab scale tests and process simulations. J. Power Sour. **206**, 393–401 (2012)

97. L. Ma, Z. Nie, X. Xi, X.g. Han, Cobalt recovery from cobalt-bearing waste in sulphuric and citric acid systems. Hydrometallurgy. **136**, 1–7 (2013)
98. E. Gratz, Q. Sa, D. Apelian, Y. Wang, A closed loop process for recycling spent lithium ion batteries. J. Power Sour. **262**, 255–262 (2014)
99. H. Zou, E. Gratz, D. Apelian, Y. Wang, A novel method to recycle mixed cathode materials for lithium ion batteries. Green Chem. **15**, 1183 (2013)
100. S. Natarajan, A.B. Boricha, H.C. Bajaj, Recovery of value-added products from cathode and anode material of spent lithium-ion batteries. Waste Management. (2018)
101. P. Meshram, B.D. Pandey, T.R. Mankhand, Hydrometallurgical processing of spent lithium ion batteries (LIBs) in the presence of a reducing agent with emphasis on kinetics of leaching. Chem. Eng. J. **281**, 418–427 (2015)
102. M. Lundström, J. Aromaa, O. Forsén, O. Hyvärinen, M.H. Barker, Leaching of chalcopyrite in cupric chloride solution. Hydrometallurgy **77**, 89–95 (2005)
103. S.P. Barik, G. Prabaharan, B. Kumar, An innovative approach to recover the metal values from spent lithium-ion batteries. Waste Manage. **51**, 222–226 (2016)
104. J.W. An, D.J. Kang, K.T. Tran, M.J. Kim, T. Lim, T. Tran, Recovery of lithium from Uyuni salar brine. Hydrometallurgy **117–118**, 64–70 (2012)
105. D. Dutta, A. Kumari, R. Panda, S. Jha, D. Gupta, S. Goel, M.K. Jha, Close loop separation process for the recovery of Co, Cu, Mn, Fe and Li from spent lithium-ion batteries. Sep. Purif. Technol. **200**, 327–334 (2018)
106. Y. Weng, S. Xu, G. Huang, C. Jiang, Synthesis and performance of Li[(Ni1/3Co1/3Mn1/3)(1-x)Mgx]O2 prepared from spent lithium ion batteries. J. Hazard. Mater. **246–247**, 163–172 (2013)
107. J. Pakarinen, E. Paatero, Recovery of manganese from iron containing sulfate solutions by precipitation. Miner. Eng. **24**, 1421–1429 (2011)
108. M. Soylak, B. Kaya, M. Tuzen, Copper(II)-8-hydroxquinoline coprecipitation system for preconcentration and separation of cobalt(II) and manganese(II) in real samples. J. Hazard. Mater. **147**, 832–837 (2007)
109. J.M. Zhao, X.Y. Shen, F.L. Deng, F.C. Wang, Y. Wu, H.Z. Liu, Synergistic extraction and separation of valuable metals from waste cathodic material of lithium ion batteries using Cyanex272 and PC-88A. Sep. Purif. Technol. **78**, 345–351 (2011)
110. T. Suzuki, T. Nakamura, Y. Inoue, M. Niinae, J. Shibata, A hydrometallurgical process for the separation of aluminum, cobalt, copper and lithium in acidic sulfate media. Sep. Purif. Technol. **98**, 396–401 (2012)
111. P. Mantuano, D. Dorella, G. Elias, R.C. A. Mansur, M. Borges, Analysis of a hydrometallurgical route to recover base metals from spent rechargeable batteries by liquid-liquid extraction with Cyanex 272. J. Power Sour. **159**, 1510–1518 (2006)
112. A.K. Jha, M.K. Jha, A. Kumari, S.K. Sahu, V. Kumar, B.D. Pandey, Selective separation and recovery of cobalt from leach liquor of discarded Li-ion batteries using thiophosphinic extractant. Sep. Purif. Technol. **104**, 160–166 (2013)
113. B. Swain, J. Jeong, J.-C. Lee, G.-H. Lee, Development of process flow sheet for recovery of high pure cobalt from sulfate leach liquor of LIB industry waste: a mathematical model correlation to predict optimum operational conditions. Sep. Purif. Technol. **63**, 360–369 (2008)
114. R. Torkaman, M. Asadollahzadeh, M. Torab-Mostaedi, M. Ghanadi Maragheh, Recovery of cobalt from spent lithium ion batteries by using acidic and basic extractants in solvent extraction process. Sep. Purif. Technol. **186**, 318–325 (2017)
115. Y. Pranolo, W. Zhang, C.Y. Cheng, Recovery of metals from spent lithium-ion battery leach solutions with a mixed solvent extractant system. Hydrometallurgy **102**, 37–42 (2010)
116. F. Pagnanelli, E. Moscardini, P. Altimari, T. Abo Atia, L. Toro, Cobalt products from real waste fractions of end of life lithium ion batteries. Waste Manage. **51**, 214–221 (2016)
117. S. Virolainen, M.F. Fini, A. Laitinen, T. Sainio, Solvent extraction fractionation of Li-ion battery leachate containing Li, Ni, and Co. Sep. Purif. Technol. **179**, 274–282 (2017)
118. Y. Yang, S. Xu, Y. He, Lithium recycling and cathode material regeneration from acid leach liquor of spent lithium-ion battery via facile co-extraction and co-precipitation processes. Waste Manage. **64**, 219–227 (2017)

119. S.-H. Joo, S.M. Shin, D. Shin, C. Oh, J.-P. Wang, Extractive separation studies of manganese from spent lithium battery leachate using mixture of PC88A and Versatic 10 acid in kerosene. Hydrometallurgy **156**, 136–141 (2015)
120. S.-H. Joo, D. Shin, C. Oh, J.-P. Wang, S.M. Shin, Extraction of manganese by alkyl monocarboxylic acid in a mixed extractant from a leaching solution of spent lithium-ion battery ternary cathodic material. J. Power Sour. **305**, 175–181 (2016)
121. Z. Jing, X. Wang, X. Peng, W. Xin, J. Huang, C. Jie, B. Louangsouphom, J. Zhao, Enhanced sunlight photocatalytic activity and recycled Ag–N co-doped TiO 2 supported by expanded graphite C/C composites for degradation of organic pollutants. Res. Chem. Intermediat **42**, 1–17 (2016)
122. K. Tanong, L.-H. Tran, G. Mercier, J.-F. Blais, Recovery of Zn (II), Mn (II), Cd (II) and Ni (II) from the unsorted spent batteries using solvent extraction, electrodeposition and precipitation methods. J. Clean. Prod. **148**, 233–244 (2017)
123. D.D.S. Leite, P.L.G. Carvalho, L.R. de Lemos, A.B. Mageste, G.D. Rodrigues, Hydrometallurgical separation of copper and cobalt from lithium-ion batteries using aqueous two-phase systems. Hydrometallurgy. **169**, 245–252 (2017)
124. A.C. Sonoc, J. Jeswiet, N. Murayama, J. Shibata, A study of the application of Donnan dialysis to the recycling of lithium ion batteries. Hydrometallurgy **175**, 133–143 (2018)
125. A. Iizuka, Y. Yamashita, H. Nagasawa, A. Yamasaki, Y. Yanagisawa, Separation of lithium and cobalt from waste lithium-ion batteries via bipolar membrane electrodialysis coupled with chelation. Sep. Purif. Technol. **113**, 33–41 (2013)
126. Q. Sa, E. Gratz, M. He, W. Lu, D. Apelian, Y. Wang, Synthesis of high performance LiNi1/3Mn1/3Co1/3O2 from lithium ion battery recovery stream. J. Power Sour. **282**, 140–145 (2015)
127. S. Zhang, C. Deng, S.Y. Yang, H. Niu, An improved carbonate co-precipitation method for the preparation of spherical Li[Ni1/3Co1/3Mn1/3]O2 cathode material. J. Alloy. Compd. **484**, 519–523 (2009)
128. F. Zhou, X. Zhao, A. van Bommel, A.W. Rowe, J.R. Dahn, Coprecipitation Synthesis of NixMn1−x(OH)2Mixed Hydroxides†. Chem. Mater. **22**, 1015–1021 (2010)
129. M. Jo, H. Ku, S. Park, J. Song, K. Kwon, Effects of Residual Lithium in the precursors of Li[Ni1/3Co1/3Mn1/3]O2 on their lithium-ion battery performance. J. Phys. Chem. Solids **118**, 47–52 (2018)
130. A.M.A. Hashem, A.E. Abdel-Ghany, A.E. Eid, J. Trottier, K. Zaghib, A. Mauger, C.M. Julien, Study of the surface modification of LiNi1/3Co1/3Mn1/3O2 cathode material for lithium ion battery. J. Power Sour. **196**, 8632–8637 (2011)
131. X. Zhang, Y. Bian, S. Xu, E. Fan, Q. Xue, Y. Guan, F. Wu, L. Li, R. Chen, Innovative application of acid leaching to regenerate Li(Ni1/3Co1/3Mn1/3)O2 cathodes from spent lithium-ion batteries. ACS Sustain. Chem. Eng. (2018)
132. L.-P. He, S.-Y. Sun, J.-G. Yu, Performance of LiNi1/3Co1/3Mn1/3O2 prepared from spent lithium-ion batteries by a carbonate co-precipitation method. Ceram. Int. **44**, 351–357 (2018)
133. C. Deng, S. Zhang, L. Ma, Y.H. Sun, S.Y. Yang, B.L. Fu, F.L. Liu, Q. Wu, Effects of precipitator on the morphological, structural and electrochemical characteristics of Li[Ni1/3Co1/3Mn1/3]O2 prepared via carbonate coprecipitation. J. Alloy. Compd. **509**, 1322–1327 (2011)
134. L. Yao, Y. Feng, G. Xi, A new method for the synthesis of LiNi1/3Co1/3Mn1/3O2 from waste lithium ion batteries. RSC Adv. **5**, 44107–44114 (2015)
135. T. Zhao, S. Chen, L. Li, X. Zhang, R. Chen, I. Belharouak, F. Wu, K. Amine, Synthesis, characterization, and electrochemistry of cathode material Li[Li0.2Co0.13Ni0.13Mn0.54]O2 using organic chelating agents for lithium-ion batteries. J. Power Sour. **228**, 206–213 (2013)
136. H. Xia, H. Wang, W. Xiao, L. Lu, M.O. Lai, Properties of LiNi1/3Co1/3Mn1/3O2 cathode material synthesized by a modified Pechini method for high-power lithium-ion batteries. J. Alloy. Compd. **480**, 696–701 (2009)
137. Z.-D. Huang, X.-M. Liu, S.-W. Oh, B. Zhang, P.-C. Ma, J.-K. Kim, Microscopically porous, interconnected single crystal LiNi1/3Co1/3Mn1/3O2 cathode material for Lithium ion batteries. J. Mater. Chem. **21**, 10777 (2011)

138. J. Li, L. Hu, H. Zhou, L. Wang, B. Zhai, S. Yang, P. Meng, R. Hu, Regenerating of LiNi0.5Co0.2Mn0.3O2 cathode materials from spent lithium-ion batteries. J. Mat. Sci. Mat. Electron. **29**, 17661–17669 (2018)
139. M. Lu, H.N. Zhang, B.C. Wang, X.D. Zheng, C.S. Dai, The re-synthesis of LiCoO2 from Spent lithium ion batteries separated by vacuum-assisted heat-treating method. Int J Electrochem Sci. **8**, 8201–8209 (2013)
140. X. Meng, H. Cao, J. Hao, P. Ning, G. Xu, Z. Sun, Sustainable preparation of LiNi1/3Co1/3Mn1/3O2–V2O5 cathode materials by recycling waste materials of spent lithium-ion battery and vanadium-bearing Slag, ACS Sustain. Chem. Eng. (2018)
141. H. Nie, L. Xu, D. Song, J. Song, X. Shi, X. Wang, L. Zhang, Z. Yuan, LiCoO2: recycling from spent batteries and regeneration with solid state synthesis. Green Chem. **17**, 1276–1280 (2015)
142. X. Song, T. Hu, C. Liang, H.L. Long, L. Zhou, W. Song, L. You, Z.S. Wu, J.W. Liu, Direct regeneration of cathode materials from spent lithium iron phosphate batteries using a solid phase sintering method. Rsc Adv. **7**, 4783–4790 (2017)
143. Z.M. Zhang, W.Z. He, G.M. Li, J. Xia, H.K. Hu, J.W. Huang, Renovation of LiCoO2 crystal structure from spent lithium ion batteries by ultrasonic hydrothermal reaction. Res. Chem. Intermediat **41**, 3367–3373 (2015)

High Value-Added Products From Recycling of Spent Lithium-Ion Batteries

Bin Huang, Guangzhe Li and Liang An

Abstract Due to the complexity and diversity of the electrode materials used in lithium-ion batteries (LIBs), the processes for recycling LIBs are much more complicated than those for recycling other batteries (e.g. lead-acid batteries); and for that reason, various products can be obtained from recycling of spent LIBs. The expected products depend not only on the materials fed into the recycling stream, but also the recycling processes. With the development of recycling technologies, more high value-added products can be obtained from spent LIBs. This chapter gives an overview of the products derived from different recycling methods and various electrode materials.

1 Introduction

LIBs have gradually seized the market share of consumer electronics (CEs) since their first commercialization in 1990s due to their unique advantages such as high energy density, long lifespan and low self-discharge [1, 2]. In the recent decades, with the rapid advances in CEs and electric vehicles (EVs), the need and production of LIBs have been increased tremendously [3, 4]. Particularly, it is noteworthy that the global EV market is beginning to boom, which will undoubtedly further accelerate the production of LIBs [5]. Like other industrial products, all the manufactured LIBs will ultimately reach the end of their lives. Although we may not face serious disposal problems of retired LIBs from EVs in a few years since the lifetime of EV batteries is much longer than the ones used in CEs and portable electric tools [4, 6], we should aware that numerous retired LIBs will be placed in recycle stations and wait for treatments.

B. Huang · G. Li · L. An (✉)
Department of Mechanical Engineering, The Hong Kong Polytechnic University, Hung Hom, Kowloon, Hong Kong SAR, China
e-mail: liang.an@polyu.edu.hk

B. Huang
College of Chemistry and Bioengineering, Guilin University of Technology, Guilin 541004, Guangxi, China

L. An (ed.), *Recycling of Spent Lithium-Ion Batteries*,
https://doi.org/10.1007/978-3-030-31834-5_6

LIBs are composed of cathodes, anodes, separators, electrolytes and outer cases, which contain hazardous substances such as fluorine-containing compounds, as well as high-value metals such as Li, Co, Ni, Cu, Al, etc. [7, 8]. After recycling processes, the metal values in spent LIBs will be transferred into other substances (e.g. alloys, slags, solutions, precipitates, etc.) [9]. For instance, at the end stage of a pyrometallurgical process the tradition metals, such as cobalt, nickel, copper and a part of iron form an alloy, which usually undergoes further separation and refinement to produce pure metals. The aluminum and lithium cannot be recovered in elemental state, but go into the slag which can be used as an aggregate in concrete. In contrast, leach liquor containing various metal ions is obtained in a hydrometallurgical process, which can be processed into not only pure metals, but also re-synthesized chemicals, electrode materials and some other functional materials. The production of high value-added products is the main purpose of a recycling process, particularly for the recycling processes aiming at high economic efficiency.

2 Pure Metals and Metal Compounds

2.1 Pure Metals

Pure metals, including cobalt, nickel and copper, have been successfully recovered from spent lithium ion batteries [10–13]. The recovery of pure metals normally starts with physical disassembling of cells, followed by a metal leaching and an electrowinning process. The metal leaching can be realized via chemical leaching by using acid or oxidative agents, or via an electrochemical leaching technique. After metal leaching, the obtained leach liquor is placed in a two-electrode electrochemical cell. Multiple metal ions in leach liquor can be easily separated from each other by the following electrowinning due to their different reduction potentials. Therefore, the targeted metal can be easily recovered after selective deposition.

The recovery of metallic cobalt from spent cellular telephone lithium ion batteries (LIBs) was first studied in 2007 [12]. The positive electrode of LIBs, containing multiple components including $LiCoO_2$, Co_3O_4, aluminum and carbon, was first leached in acid to obtain the electrodeposition solution, followed by electrodeposition under potentiostatic and potentiodynamic conditions. The electrodeposition of cobalt started at the potential of −0.8 V versus reference electrode (Ag/AgCl). It was found that the grain size of cobalt is highly dependent on the pH value of the leaching liquor. When the pH value is 5.40, a large grain size of electrodeposits with micrometer-scale was obtained; with a more acidic condition (pH = 2.70), the grain size of cobalt was decreased to a nanometer-scale and the amount of grains became larger. The composition of surface deposition layer was demonstrated to be 100% cobalt, without any impurities. Irregular deposition of cobalt would occur when further decreasing pH value to 1.50 and hydrogen gas would generate due to the reduction of H^+, resulting in the formation of metal layers with holes. Therefore, metallic cobalt with high purities could be recovered from spent LIBs by optimizing the pH value of liquor.

A more efficient method for metallic cobalt recovery was reported recently by employing the supercritical carbon dioxide extraction technique in metal leaching [10]. The advantages of employing such technique can be summarized as follows: Firstly, supercritical CO_2 can enable faster reaction kinetics to leach metal more efficiently. Secondly, supercritical CO_2 together with solvent can extract cobalt from the substrate more thoroughly. Lastly, the amount of solvent in use can be reduced. As a result, 98% extraction of cobalt was achieved only in a short reaction time of 5 min and with the usage of 4% (v/v) H_2O_2. However, as for the conventional technique, a reaction time of 60 min and the usage of 8% (v/v) H_2O_2 were required to achieve the same efficiency of cobalt extraction.

In another work, metallic cobalt, metallic copper and manganese dioxides were successfully recovered via typical metal leaching and electrowinning method [13]. The complete reaction time for all metals in leaching process was about 150 min and the optimized pH value of electrodeposition solution for subsequent electrowinning was in the range of 2 and 2.5. The recovery of Co, Cu and Mn from spent LIBs via above strategy was over 96, 97 and 99%, respectively. The feasibility of this technique for achieving practical applications was also analyzed based on the economic data. It was found that with the input of per kilogram electrode material, the cost was estimated about 48 Rs for the usage of sulphuric acid, caustic and the power. The metal and chemical recovered from the input were 0.013 kg Cu, 0.21 kg Co and 0.172 kg MnO_2, with an output value about 755 Rs.

Lithium/cobalt/nickel oxide is used as the commercial cathode material in LIBs. Metallic nickel was hard to be recovered independently from the lithium/cobalt/nickel oxide due to the co-deposition of nickel and cobalt in the electrowinning process. Therefore, separation of nickel and cobalt is required before the electrowinning starts, making the recovery technique rather complex. In this consideration, the independent recovery of metallic nickel was reported by using saponification agent in the metal leaching process [11]. Specifically, 0.5 M saponified CRANEX 272 was added to the leaching solution to dramatically enhance the separation effect between nickel and cobalt. The cobalt anomalous deposition was thus hindered in the following electrowinning. After a reaction time of 80 min, most of nickel in the electrodeposition solution (1.7–1.8 g/L) was deposited, leaving less than 100 ppm of nickel in the electrodeposition solution.

2.2 *Metal Compounds*

The recycling of metal compounds from spent LIBs is extremely important in modern industry for several reasons. Firstly, the metal resources in the earth crust are limited and overuse of metal resources in modern industry will cause resource depletion. Secondly, the waste from the spent LIBs will create serious environmental pollution if no recycling measures are carried out. Thirdly, the recycled metal compounds can be directly used in the synthesis of electrode materials, showing values from an economic point of view.

2.2.1 Lithium Compounds

The recovery of lithium compounds gains lots of attractions due to the high expense of lithium compounds. In most lithium recycling studies, lithium resources were recovered as the form of lithium carbonate (Li_2CO_3) [14]. Recently, a facile method using chemical additive was employed to recover lithium from spent LIBs [15]. By using additive (D2EHPA), lithium was well separated from other metal elements such as cobalt, nickel and manganese. As a result, lithium in the leach liquor was recovered to Li_2CO_3 with a high purity of 99.2% after precipitation. Hu et al. [16] reported a promising approach to obtain Li_2CO_3 by carbonated water leaching. The procedure of the recycling was shown in Fig. 1. It began with the alkaline leaching, in which cathode materials were converted into mixture powder. A roasting process at 650 °C was employed to reduce the residue into metal, metal oxide and metal carbonate. Owing to the low solubility of Li_2CO_3 in water, carbonated water was used to react with Li_2CO_3 and transformed Li_2CO_3 into $LiHCO_3$ with better water solubility. The $LiHCO_3$ solution was then evaporated and Li_2CO_3 was obtained after evaporation crystallization. The purification of Li_2CO_3 in roasted products was thus achieved. This environmentally friendly method was able to recover 84.7% lithium from the spent cathode materials.

Lithium resources of spent LIBs can also be recovered as the form of Li_3PO_4 [17]. Specifically, the hybrid cathode powder, mainly containing $LiFePO_4$ and $LiMn_2O_4$, was first leached in acid and converted into metal-enriched solution. After progressive purification, lithium-enriched solution was obtained. By adding Na_3PO_4, lithium was finally precipitated as Li_3PO_4.

2.2.2 Manganese Compounds

Manganese is one of the wide spread elements in the earth crust and is also an essential component in batteries. Manganese oxides like MnO_2 or Mn_2O_3 can be recovered from the spent LIBs with $LiMn_2O_4$ or NCM cathode. For example, MnO_2 was successfully recovered from the powder mixture containing $LiCoO_2$, $LiMn_2O_4$ and $LiCo_{1/3}Ni_{1/3}Mn_{1/3}O_2$ with 1:1:1 weight ratio [18]. By leaching spent electrode material mixture in acid solution, the leach liquor containing manganese was obtained. The precipitation of manganese began at a pH value of 1. After adding potassium permanganate reagent drop by drop, the following reaction would happen:

$$3Mn^{2+} + 2MnO_4^- + 2H_2O \rightarrow 5MnO_2 + 4H^+ \quad (1)$$

The precipitation of manganese completed at a pH value of 2 and the purity of obtained MnO_2 product was 98.23%. Similarly, the recovery of MnO_2/Mn_2O_3

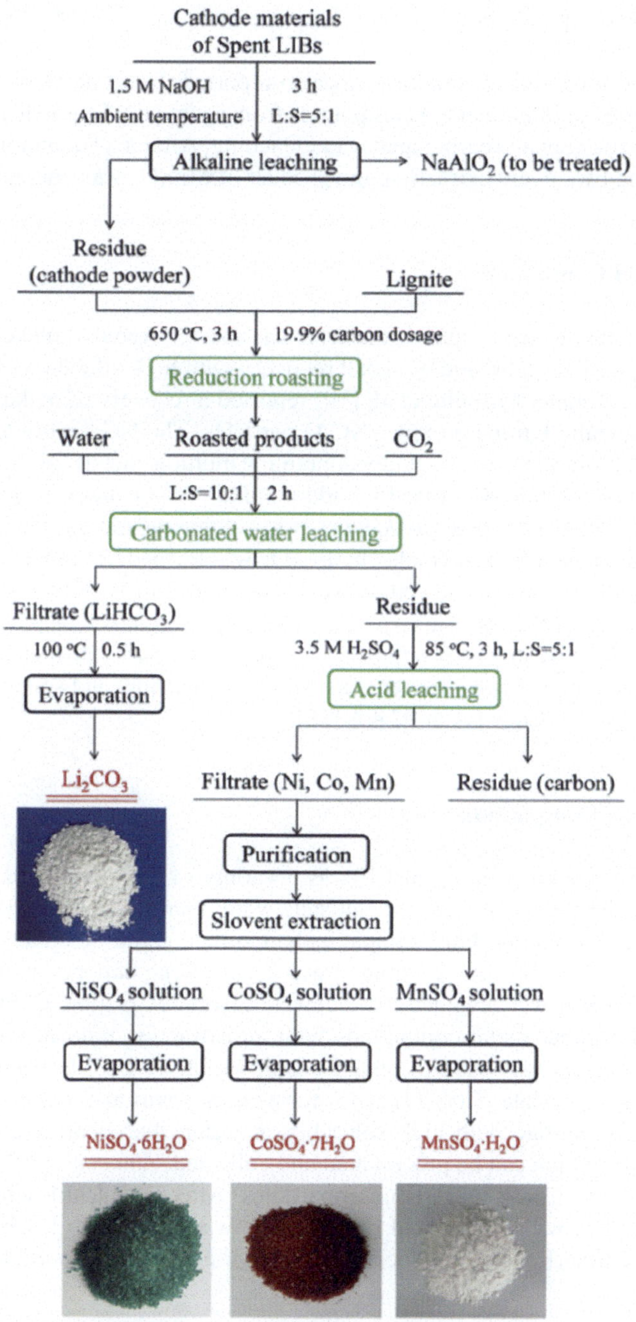

Fig. 1 Process flow diagram for the recovery of valuable chemicals from spent LIBs, Ref. [16]. Reproduced with permission from Elsevier

mixture was also reported by adding potassium permanganate and Mn^{2+} was oxidized to multiple valence [17].

The recovery of $MnSO_4$ product was also reported. Hu et al. [16] obtained the residue mixture containing metal compounds from spent NCM cathode materials by the reduction roasting and carbonated water leaching. After acid treatment of residue and separation by solvent extraction, the product of $MnSO_4$ was obtained.

2.2.3 Nickel Compounds

Nickel component in spent cathode material was also recovered as nickel compound for specific use. The technique of nickel recovery with high efficiency was reported recently. For example, M.Joulie et al. [19] reported a recovery of nickel compound from nickel cobalt aluminum oxide (NCA) particles. The NCA particles were first leached in acid solution to obtain liquor containing multiple metal ions. Subsequently, the pH value of solution was tuned by adding alkaline to extract Co ions from the liquor. The Co ions were precipitated as Co_2O_3. After separation, the alkaline was further added into solution and nickel hydroxide was formed. A high nickel recovery efficiency of 99.99% was achieved, however, the purity of $Ni(OH)_2$ was relatively low, with a purity of 96.36%. This was due to incomplete removal of cobalt in the first precipitation. Post treatment is required to improve the purity of nickel hydroxide. In another study, nickel was recovered as $NiSO_4$ by leaching and solvent extraction with a recovery efficiency about 97.4% [16].

2.2.4 Cobalt Compounds

Cobalt was recovered from spent LIBs as a variety of compounds such as cobalt salts and cobalt oxides. Different cobalt salts were obtained by using various types of acid in metal leaching. For example, cobalt sulfate could be obtained by sulfur acid leaching [16]. After acid leaching of waste, a leaching liquor containing $NiSO_4$, $CoSO_4$ and $MnSO_4$ was obtained. By stepwise solvent extraction, $CoSO_4$ was well separated from other metal compounds with an extraction ratio of 98.2%. Direct solvent evaporation was used to obtain cobalt sulfate due to its water solubility. Other metal salt, oxalate ($CoC_2O_4 \cdot H_2O$) for example, was also reported. Different from the cobalt sulfate with high solubility in water, the cobalt oxalate shows a poor solubility, which can be precipitated and collected. Zhu et al. [20] reported the recovery of Co as $CoC_2O_4 \cdot 2H_2O$ microparticles. After chemical leaching of spent cathode materials with sulfur acid $(NH_4)_2C_2O_4$ was added into leach liquor. The following reaction (Eq. 2) would occur and $CoC_2O_4 \cdot 2H_2O$ microparticles were thus precipitated.

$$CoSO_4(aq) + (NH_4)_2C_2O_4(aq) \rightarrow CoC_2O_4(s) \downarrow + (NH_4)_2SO_4(aq) \quad (2)$$

As a result, 94.7% cobalt was recovered and the impurity of $CoC_2O_4{\cdot}2H_2O$ was less than 0.32%. Similarly, Chen et al. [21] reported the recovery of CoC_2O_4 with higher efficiency by adding excess of $(NH_4)_2C_2O_4$. As a result, over 95% of Co was recovered from spent LIBs and the purity of final product CoC_2O_4 was more than 99%. Except cobalt oxalate, the recovery of cobalt as cobalt phosphate was also reported by chen et al. [22] More than 99% Co was extracted by chemical leaching. By adding mild phosphoric acid, the precipitation of cobalt phosphate would occur and after filtration, the obtained product had a high purity of 97.1%.

Cobalt recovered as the carbonate was also reported [23]. Saturated Na_2CO_3 solution was used to purify the leach liquors under stirring and to final pH value of solution was in the range of 9–10. In the process of stirring, the cobalt ions would react with Na_2CO_3 and deposit as cobalt carbonate ($CoCO_3$). After washing with water to remove the soluble sodium sulfate, the purity of the obtained $CoCO_3$ was 47% (w/w of cobalt). Wang et al. [18] reported the recovery of Co as the cobalt hydroxide $Co(OH)_2$. Ammonia solution was first used to extract nickel ions from leach liquor. After the removal of nickel, the excess of ammonia ions and cobalt ions would exist in the solution. Since the hydroxide would partially dissolute into the solution with the presence of $[Co(NH_3)_6]^{3+}$, an acid condition was required before the precipitation of cobalt hydroxide. In this consideration, hydrochloric acid was first added into the leach liquor to adjust the pH value of solution to 0 to avoid the dissolution of cobalt hydroxide. Then, NaOH was added into solution to induce the precipitation of $Co(OH)_2$. The final product had a high purity of 96.94%.

Apart from cobalt salts, cobalt oxides were also recovered from the spent cathode materials. The recycling products could be reused in the direct synthesis of cathode materials, building up a closed loop in LIB industry. For example, Hu et al. [24] reported the preparation of nanoscaled cobalt oxide (Co_3O_4) by using recycled cobalt oxalate as a precursor. The preparation contained the following steps: Firstly, the active powder mixture was leached by chemicals to obtain the leach liquor. Second, by adding alkaline, impurity ions were precipitated in turn by adjusting pH value of the solution, leaving cobalt ions only. Thirdly, the CoC_2O_4 products were precipitated from the leach liquor by adding $(NH_4)_2C_2O_4$ saturation solution into leaching liquor. Fourthly, the collected CoC_2O_4 powder was dissolved in HNO_3 acid. After adding citric acid, the gel was formed at a temperature of 70–80 °C. Lastly, the gel was transferred into a container and was heated in the air to obtain Co_3O_4 product. The obtained product showed a structure of aggregated nanoparticles and the average diameter of these particles was about 80 nm. When it used as the anode material in LIBs, an initial discharge capacity of 760.9 mAh g^{-1} was delivered, and the reversible capacity was about 442.3 mAh g^{-1} at a current density of 250 mA g^{-1} after 20 cycles. In another work, $Co_2O_3.3H_2O$ was recovered from the spent lithium nickel cobalt aluminum oxide (NCA) cathodes [19]. After chemical leaching, cobalt compound was converted into Co (II) in the liquor. By using NaClO reagent, the Co (II) in the liquor was selectively oxidized into Co (III), following the below reaction:

$$2Co^{2+} + ClO^- + 2H_3O^+ \rightarrow 2Co^{3+} + Cl^- + 3H_2O \quad (3)$$

After oxidization, the pH of solution was then tuned to a value of 3 to precipitate the cobalt ions in acid media as the following reaction:

$$2Co^{3+} + 6HO^{-} \rightarrow Co_2O_3 \cdot 3H_2O \quad (4)$$

The purity of cobalt hydroxide product was 90.25 wt%, showing good recovery efficiency.

2.2.5 Ferrum Compounds

Owing to the industrialization success of $LiFePO_4$ cathode materials in recent years, the consumption of ferrum grows quickly in the battery industry. Therefore, it is worthy to save ferrum resources by effective recycling. Huang et al. [17] reported the recovery of ferrum as ferric chloride ($FeCl_3$). After the chemical leaching of spent $LiFePO_4$ cathode materials, lithium ions and ferrum ions coexisted in the leach liquor. The recovery efficiency of this step was about 85.4% and the product could be reused in the preparation of $LiFePO_4$ cathode materials, which brings huge benefits both economically and environmentally.

3 Electrode Materials

Various pure metals and metal compounds could be recovered from the spent LIBs, showing multiple functionalities and values in modern industries. Specifically, in the battery industry, a closed loop to regenerate cathode materials from the spent LIBs will promote the formation of industrial chain and thus be more efficient. An overview on the regeneration of cathode materials by using spent cathode materials as precursors was given as follows.

3.1 $LiCoO_2$ Cathode Material

$LiCoO_2$ active material was successfully recovered from spent LIBs [25]. After the mechanical separation to disassemble the battery package and thermal treatment to burn off the binder and carbon, the active $LiCoO_2$ powder was obtained, with an average particle size of 15 μm. The leaching process was carried out in HNO_3 solution to dissolve lithium and cobalt, followed by a sol-gel process with the presence of citric acid. After the heat treatment of gel precursor in air, the regeneration of $LiCoO_2$ was completed. The lithium storage properties of newly generated $LiCoO_2$ were also studied, showing initial discharge capacity and charge capacity of 154 mAh g^{-1} and 165 mAh g^{-1}, respectively. The capacity retention was about 93% of the initial capacity after 30 cycles.

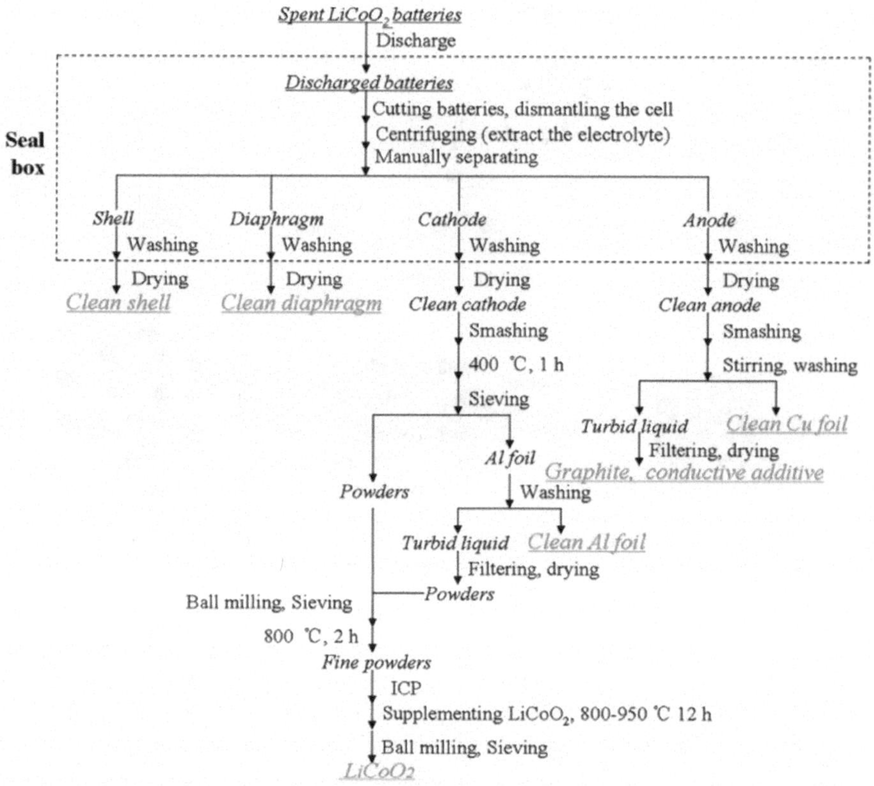

Fig. 2 A diagram of the recycling process and the following synthesis of $LiCoO_2$ active material, Ref. [26]. Reproduced with permission from ROYAL SOCIETY OF CHEMISTRY

Nie et al. [26] reported the solid state synthesis of $LiCoO_2$ by recycling spent $LiCoO_2$ batteries. The procedure of the recycling and synthesis is shown in Fig. 2. In addition to the $LiCoO_2$ product, the clean shell, clean diaphragm, clean Al foil, clean Cu foil, conductive additive and graphite were also recycled simultaneously. The resultant active material exhibited a reversible capacity of 152.4 mAh g^{-1} and good cycling stability.

The re-synthesis of $LiCoO_2$ was also reported in other studies, showing good recovery efficiency as well as lithium storage properties [27–30].

3.2 *Li(Co–Mn–Ni)O_2 Cathode*

The recovery of NCM cathode material from spent LIBs are numerously reported recently [31–37]. For example, $LiCo_{0.415}Mn_{0.435}Ni_{0.15}O_2$ was re-synthesized from the spent LIBs by co-precipitation and heat treatment [37]. A lithium storage capacity of 64 mAh g^{-1} was achieved at a high current density of 40 C. Besides, the

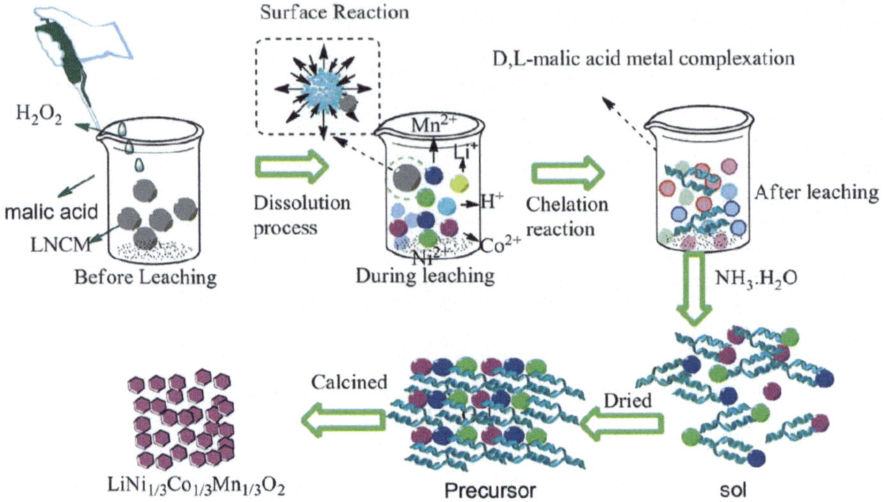

Fig. 3 A schematic illustration of recycling of spent LIBs and regeneration of $LiNi_{1/3}Co_{1/3}Mn_{1/3}O_2$ active material, Ref. [36]. Reproduced with permission from RSC Publishing

active material also showed good sodium ion storage properties, with a capacity of 93 mAh g^{-1} at a current density of 100 mA g^{-1}. The recovery of $LiNi_{1/3}Co_{1/3}Mn_{1/3}O_2$ was also reported by Yao et al. [36]. A new process for recycling $LiNi_{1/3}Co_{1/3}Mn_{1/3}O_2$ was employed, as shown in Fig. 3. After initial chemical leaching of spent cathode materials, D,L-malic acid was used as both leaching reagent and chelating agent, to transfer the liquor into sol with the presence of ammonia solution. After the calcination, the final product of $LiNi_{1/3}Co_{1/3}Mn_{1/3}O_2$ was obtained, showing an initial discharge capacity of 147.2 mAh g^{-1} in the voltage range from 2.75 to 4.25 V. The capacity retention was about 95% after 100 cycles, presenting good cyclic performance.

3.3 *$LiFePO_4$ Cathode*

The recovery of $LiFePO_4$ (LFP) was also studied in detail. A novel process for synthesizing $LiFePO_4$/C microflower from spent LFP cathodes was reported recently [38]. The waste was first leached by using phosphoric acid, followed by filtration to obtain $FePO_4 \cdot 2H_2O$ precursor. After carbothermal reduction, the $LiFePO_4$/C was obtained, which owned a flower-like structure with a diameter of 1-2 micrometers. The resultant material exhibited an excellent discharge capacity of 105 mAh g^{-1} after 500 cycles at 5 C.

4 Other Functional Materials

Other compounds possessing special functions can be obtained from the recycling of spent LIBs, such as magnetic materials (e.g. NiCo ferrite, CuCo ferrite, etc.), electrocatalytic materials (e.g. spinel $MnCo_2O_4$), photocatalytic materials (e.g. $Co_3O_4/LiCoO_2$), etc.

4.1 Magnetic Materials

A survey carried out in 2012 suggested that more than half of the commercial cathode materials were cobalt-rich materials, such as $LiCoO_2$ and $LiNi_{0.33}Co_{0.33}Mn_{0.33}O_2$ [39]. Although the cobalt content in cathode materials is decreasing with the development of material technology, it is still essential for the electrochemical stability of the layered-structured cathode materials [40, 41]. Hence, it is expected that cobalt will still exist in spent LIBs for a long time. Other than the re-synthesized electrode materials, the cobalt-containing waste materials can be processed into cobalt ferrites, which are a common kind of magnetic material possessing the potential of versatile applications, such as information storage, gas sensing, stress sensing and non-contact sensing [42–44].

Generally, cobalt ferrites are synthesized using the leach liquor of waste cathode materials. Many methods to synthesize cobalt ferrites have been investigated and reported, such as sol-gel methods [45, 46], hydrothermal methods [47, 48] and co-precipitation methods [49, 50]. The microstructures and crystalline sizes of the cobalt ferrites could have significant impacts on the magnetic properties [51]. In other words, the magnetic properties could be altered by employing different synthetic methods or regulate the synthesis conditions. For instance, Yao et al. [45] adopted a sol-gel-hydrothermal method to synthesize a $CoFe_2O_4$ precursor which has a hedgehog-like spherical morphology with the dimension of around 5 μm using leach liquor of spent $LiCoO_2$ cathode material as raw material. In a typical synthesis, the collected waste cathode material powder was dissolved in 3.5 mol L^{-1} H_2SO_4 and 10% H_2O_2 with a pulp density of 50 g L^{-1} to produce a leach liquor containing Co^{2+} and $SO_4{}^{2-}$. Then appropriate amounts of $Fe_2(SO_4)_3$ and $CoSO_4$ were added to the leach liquor to adjust the mole ratio of Co^{2+} and Fe^{3+} to 1:2, and the total concentration of the metals to 1 mol L^{-1}. After that, citric acid was added to the solution with the mole ratio of citric acid to total metal ions in 1:1. Finally, deionized water was added to the solution and the pH value was adjusted to 8 using ammonia, followed by hydrothermal treatment at 240 °C for 1, 2, 6, 12 and 24 h. The growth of the precursor can be illustrated by the schematic diagram shown in Fig. 4.

For the characterization of magnetic properties, the precursor was subsequently mixed with polyvinyl alcohol (PVA, 8–10%) and compacted into cylindrical pellets with a diameter of 10 mm and a length of 20 mm. The pellets made of the precursor

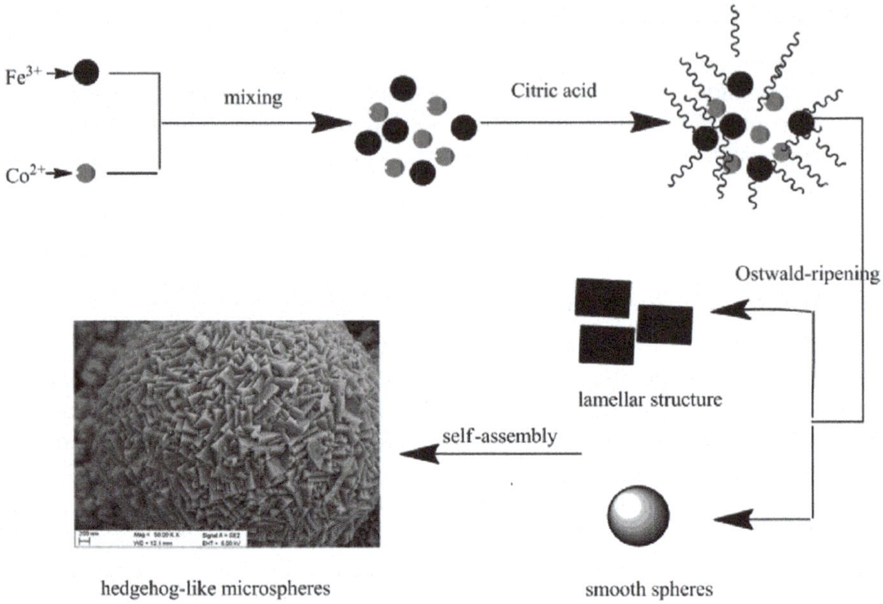

Fig. 4 Schematic illustration of the growth process of the precursor particles, Ref. [45]. Reproduced with permission from Elsevier

synthesized by hydrothermal treatment at 240 °C for 12 h exhibited the highest magnetostriction coefficient of −158.5 ppm and the highest train derivative coefficient of -1.69×10^{-9} $A^{-1}m$, as shown in Fig. 5.

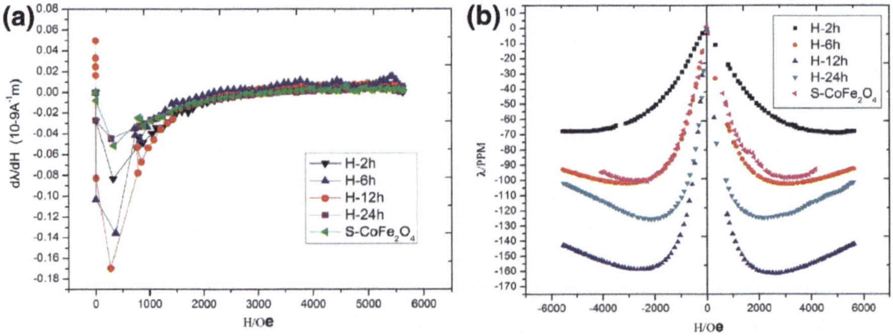

Fig. 5 **a** Strain derivative curves and **b** magnetostriction curves of the as-synthesized precursors reported in Ref. [45]. Reproduced with permission from Elsevier

4.2 *Electrocatalytic Materials*

Various iron-, cobalt-, manganese- and nickel-containing compounds, including oxides, hydroxides, sulfides, selenides and phosphides have been studied for electrocatalytic water splitting, including the hydrogen evolution reaction (HER) and oxygen evolution reaction (OER) [52–55]. The cathode materials of LIBs contain one or more such transition metals hence the spent LIBs are likely to be a source for the synthesis of the electrocatalysts. It is worth mentioning that some electrode materials also have electrocatalytic activity, such as $LiCoO_2$ and Co_3O_4. However, the methods for synthesizing these materials might depend on the specific application since an electrode material and an electrocatalyst may have different demand in crystal form, microstructure and morphology.

For obtaining electrocatalysts from spent LIBs, the waste cathode materials are usually dissolved in acidic solution to form a leach liquor firstly, then followed by the adjustments for metal-ions ratio and concentrations. Take the spinel $MnCo_2O_4$ as an example, a typical synthetic method proposed by Natarajan et al. [56] can be briefly described as following. Firstly, the cathode pieces collected after dismantling of spent LIBs were crushed and immersed into a solution of 2 M acetic acid at 80 °C for 80 min. Secondly, the leach liquor was purified and the metal ions in it were adjusted to the required ratio and concentrations. Thirdly, ammonium carbonate solution was added drop by drop to the metal acetate solution with constant stirring to precipitate the metal ions, followed by an extra constant stirring at room temperature for 24 h. Fourthly, the suspension was treated in a hydrothermal process at 150 °C for 24 h, and then the precipitate was collected and washed by methanol. Finally, the precipitate was calcined at 650 °C for 4 h under air atmosphere to produce $MnCo_2O_4$ powder.

The resultant product was spherical microspheres with the diameter of 0.5–3.0 μm, as shown in Fig. 6a. Its electrocatalytic performance was evaluated using the OER in 1 mol L^{-1} KOH solution. With the electrocatalyst loading of

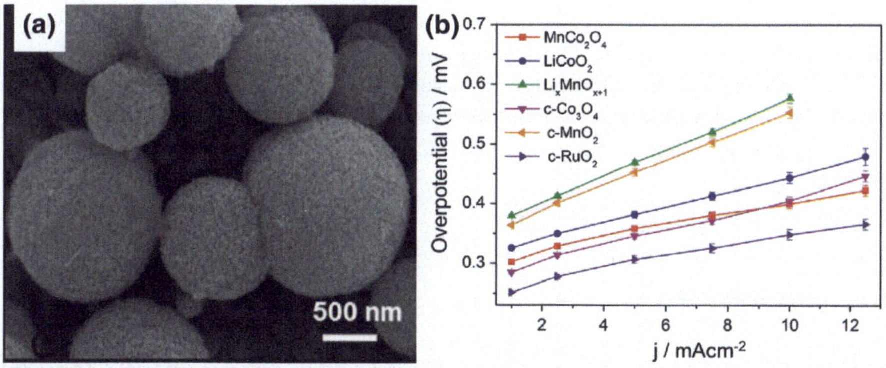

Fig. 6 **a** SEM image of the spinel $MnCo_2O_4$ and **b** over-potential-current density plots of the spinel $MnCo_2O_4$ together with other electrocatalysts samples, reported in Ref. [56]. Reproduced with permission from ROYAL SOCIETY OF CHEMISTRY

0.001025 g cm^{-2}, the OER generated current densities of 5 and 10 mA cm^{-2} with the over-potentials of 358 and 400 mV, respectively, as shown in Fig. 6b. In addition, the electrocatalytic activity of the as-prepared $MnCo_2O_4$ powder was compared with $LiCoO_2$ and Li_xMnO_{x+1} re-synthesized from spent LIBs, as well as commercial Co_3O_4, MnO_2, and RuO_2. The result, which is also depicted in Fig. 6b, showed that the $MnCo_2O_4$ was superior to all these catalysts except for the commercial RuO_2.

4.3 Photocatalytic Materials

Many transition metal oxides have photocatalytic activity and some of them containing cobalt, nickel and (or) lithium can be synthesized from spent LIBs. For example, $LiCoO_2$, one of the most common cathode materials for LIBs, can be converted to an electrocatalytic material by transforming the crystal structure from hexagonal to spinel. Recently, it has been proved that the spinel $LiCoO_2$ also had photocatalytic activity [57]. The spinel $LiCoO_2$ can be obtained by a low-temperature calcination (below 500 °C), whereas the hexagonal one must be synthesized above 500 °C [58]. It was reported by Santana et al. that the leach liquor of waste $LiCoO_2$ cathode material could be used to synthesize photocatalyst $Co_3O_4/LiCoO_2$ composite, which could discolor methylene blue dye. The synthesis and performance measurement are briefly introduced as following.

Firstly, the collected waste $LiCoO_2$ cathode material was immersed into a leaching agent which was composed of 2.0 mol L^{-1} of citric acid and 30 v/v% of H_2O_2 at 80 °C for 90 min. Secondly, 50 mL of the leach liquor was heated at 85 °C and kept in constant stirring for 5 h to form a pink sol, followed by drying at 120 °C for 24 h to form a gel. Finally, the gel was calcined at 450 °C for 3 h to form the product, $Co_3O_4/LiCoO_2$ mixture.

The obtained material possessed porous morphology, and was composed of agglomerated nano-particles, as shown in Fig. 7a. A solution of 3 mg L^{-1} methylene blue dye was employed as visual indicator to evaluate the photocatalytic activity of the $Co_3O_4/LiCoO_2$ material. The discoloration of methylene blue dye was measured by a UV-visible spectrophotometer at wavelengths of 613 and 664 nm. Figure 7b shows the discoloration rate for the absorption peaks at the adopted wavelengths. After 10 h and 24 h photocatalyzing, the discoloration efficiencies are 90 and 100%, respectively.

4.4 Adsorbents

In addition, some other functional materials can be made from spent LIBs. In recent years, some researchers tried to convert waste graphite anode materials into carbonaceous adsorbents due to their unique microstructure and abundant surface functional

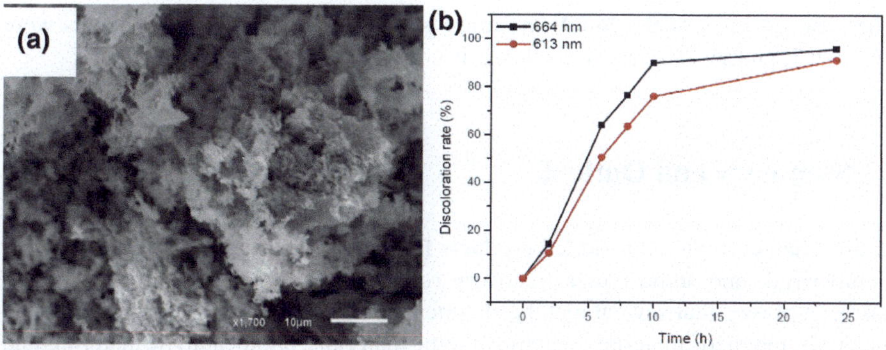

Fig. 7 **a** SEM image of the $Co_3O_4/LiCoO_2$, and **b** discoloration rate of methylene blue dye for absorption peak at 664 nm and 613 nm, reported in Ref. [57]. Reproduced with permission from Elsevier

groups. In addition, it has been proved that some manganese oxides are good adsorbents for the treatment of heavy-metal-contaminated water, hence some manganese-based cathode materials can also be made into adsorbents.

In 2017, Zhao et al. [59] firstly reported a MnO_2-modified graphite adsorbent synthesized from spent LIBs, as shown in Fig. 8a. In their study, the pure graphite adsorbent was obtained by calcining the waste graphite anode powder under nitrogen atmosphere at 600 °C for 1 h; the MnO_2-modified graphite was prepared by simply mixing the pure graphite adsorbent with 0.4 mol L^{-1} $KMnO_4$ solution (pH = 2) under constant stirring at 60 °C for 4 h, followed by washing and drying at 80 °C. Both the recovered pure graphite and the MnO_2-modified graphite show adsorbability to Pb(II), Cd(II), and Ag(I), but the latter showed much improved removal capability.

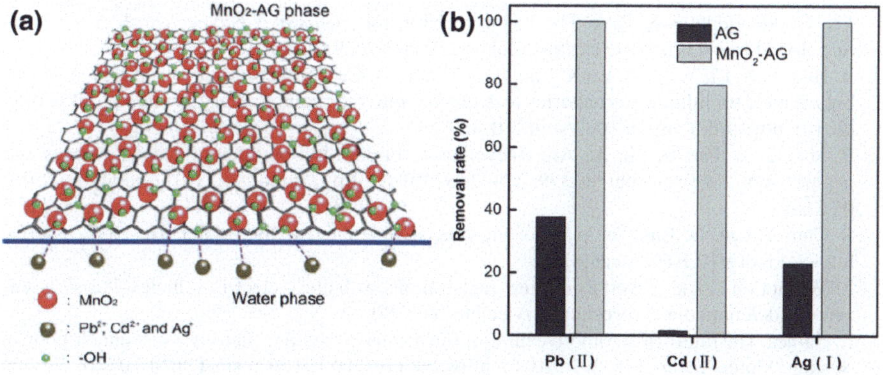

Fig. 8 **a** Schematic diagram of the MnO_2-modified graphite, and **b** the comparison of the removal rates of the pure graphite and MnO_2-modified graphite to Pb(II), Cd(II) and Ag(I), reported in Ref. [59]. Reproduced with permission from Copyright (2017) American Chemical Society

The removal rates of the MnO_2-modified graphite to the three heavy metals were 99.9, 79.7, and 99.8%, respectively, as shown in Fig. 8b.

5 Summary and Outlook

In this chapter, high value-added products from recycling of spent LIBs have been summarized, and some typical recovery routes have been discussed in detail. It has been proved that several metals, chemical compounds, re-synthesized electrode materials, magnetic materials, electrocatalytic materials, photocatalytic materials and adsorbents can be obtained from the recycling of spent LIBs. However, preparing the high value-added products requires complicated treatments and expensive reagents, making the recycling processes less economical. For facilitating the recycling of spent LIBs and recovering high value-added products, more efforts for reducing the complexity of the recycling processes and enhancing the recycling efficiency are needed. For example, the spent LIBs could be identified and sorted before recycling so that different types of waste materials would not be blended. As a result, the recycling processes could be simpler and the impurities in the final products might be less.

Acknowledgements This work was fully supported by a grant from the Research Grants Council of the Hong Kong Special Administrative Region, China (Project No. 15222018).

References

1. J.B. Goodenough, K.S. Park, The Li-ion rechargeable battery: A perspective. J. Am. Chem. Soc. **135**, 1167–1176 (2013). https://doi.org/10.1021/ja3091438
2. B. Huang, X. Li, Y. Pei, S. Li, X. Cao, R.C. Massé, G. Cao, Novel carbon-encapsulated porous SnO2 anode for lithium-ion batteries with much improved cyclic stability. Small **12**, 1945–1955 (2016). https://doi.org/10.1002/smll.201503419
3. B. Huang, Z. Pan, X. Su, L. An, Recycling of lithium-ion batteries: Recent advances and perspectives. J. Power Sources **399**, 274–286 (2018). https://doi.org/10.1016/j.jpowsour.2018.07.116
4. S. Chu, Y. Cui, N. Liu, The path towards sustainable energy. Nat. Mater. **16**, 16–22 (2016). https://doi.org/10.1038/nmat4834
5. ZSW data on stock as well as on new registrations of battery-electric vehicles. https://www.zsw-bw.de/en/media-center/data-service.html#c6700
6. L. Gaines, The future of automotive lithium-ion battery recycling: charting a sustainable course. Sustain. Mater. Technol. **1**, 2–7 (2014). https://doi.org/10.1016/j.susmat.2014.10.001
7. X. Chen, C. Luo, J. Zhang, J. Kong, T. Zhou, Sustainable recovery of metals from spent lithium-ion batteries: a green process. ACS Sustain. Chem. Eng. **3**, 3104–3113 (2015). https://doi.org/10.1021/acssuschemeng.5b01000
8. X. Zeng, J. Li, L. Liu, Solving spent lithium-ion battery problems in China: opportunities and challenges. Renew. Sustain. Energy Rev. **52**, 1759–1767 (2015). https://doi.org/10.1016/j.rser.2015.08.014

9. A. Chagnes, B. Pospiech, A brief review on hydrometallurgical technologies for recycling spent lithium-ion batteries. J. Chem. Technol. Biotechnol. **88**, 1191–1199 (2013). https://doi.org/10.1002/jctb.4053
10. D.A. Bertuol, C.M. Machado, M.L. Silva, C.O. Calgaro, G.L. Dotto, E.H. Tanabe, Recovery of cobalt from spent lithium-ion batteries using supercritical carbon dioxide extraction. Waste Manage. **51**, 245–251 (2016). https://doi.org/10.1016/j.wasman.2016.03.009
11. C. Lupi, M. Pasquali, Electrolytic nickel recovery from lithium-ion batteries. Miner. Eng. **16**, 537–542 (2003). https://doi.org/10.1016/S0892-6875(03)00080-3
12. M.B.J.G. Freitas, E.M. Garcia, Electrochemical recycling of cobalt from cathodes of spent lithium-ion batteries. J. Power Sources **171**, 953–959 (2007). https://doi.org/10.1016/j.jpowsour.2007.07.002
13. G. Prabaharan, S.P. Barik, N. Kumar, L. Kumar, Electrochemical process for electrode material of spent lithium ion batteries. Waste Manag **68**, 527–533 (2017). https://doi.org/10.1016/j.wasman.2017.07.007
14. W. Gao, X. Zhang, X. Zheng, X. Lin, H. Cao, Y. Zhang, Z. Sun, Lithium carbonate recovery from cathode scrap of spent lithium-ion battery: a closed-loop process. Environ. Sci. Technol. **51**, 1662–1669 (2017). https://doi.org/10.1021/acs.est.6b03320
15. Y. Yang, S. Xu, Y. He, Lithium recycling and cathode material regeneration from acid leach liquor of spent lithium-ion battery via facile co-extraction and co-precipitation processes. Waste Manage. **64**, 219–227 (2017). https://doi.org/10.1016/j.wasman.2017.03.018
16. J. Hu, J. Zhang, H. Li, Y. Chen, C. Wang, A promising approach for the recovery of high value-added metals from spent lithium-ion batteries. J. Power Sources **351**, 192–199 (2017). https://doi.org/10.1016/j.jpowsour.2017.03.09
17. Y. Huang, G. Han, J. Liu, W. Chai, W. Wang, S. Yang, S. Su, A stepwise recovery of metals from hybrid cathodes of spent Li-ion batteries with leaching-flotation-precipitation process. J. Power Sources **325**, 555–564 (2016). https://doi.org/10.1016/J.JPOWSOUR.2016.06.072
18. R.C. Wang, Y.C. Lin, S.H. Wu, A novel recovery process of metal values from the cathode active materials of the lithium-ion secondary batteries. Hydrometallurgy **99**, 194–201 (2009). https://doi.org/10.1016/j.hydromet.2009.08.005
19. M. Joulié, R. Laucournet, E. Billy, Hydrometallurgical process for the recovery of high value metals from spent lithium nickel cobalt aluminum oxide based lithium-ion batteries. J. Power Sources **247**, 551–555 (2014). https://doi.org/10.1016/j.jpowsour.2013.08.128
20. S.G. Zhu, W.Z. He, G.M. Li, X. Zhou, X.J. Zhang, J.W. Huang, Recovery of Co and Li from spent lithium-ion batteries by combination method of acid leaching and chemical precipitation. Trans. Nonferrous Met. Soc. China (English Ed). **22**, 2274–2281 (2012). https://doi.org/10.1016/s1003-6326(11)61460-x
21. L. Chen, X. Tang, Y. Zhang, L. Li, Z. Zeng, Y. Zhang, Process for the recovery of cobalt oxalate from spent lithium-ion batteries. Hydrometallurgy **108**, 80–86 (2011). https://doi.org/10.1016/j.hydromet.2011.02.010
22. X. Chen, H. Ma, C. Luo, T. Zhou, Recovery of valuable metals from waste cathode materials of spent lithium-ion batteries using mild phosphoric acid. J. Hazard. Mater. **326**, 77–86 (2017). https://doi.org/10.1016/j.jhazmat.2016.12.021
23. G. Granata, E. Moscardini, F. Pagnanelli, F. Trabucco, L. Toro, Product recovery from Li-ion battery wastes coming from an industrial pre-treatment plant: Lab scale tests and process simulations. J. Power Sources **206**, 393–401 (2012). https://doi.org/10.1016/j.jpowsour.2012.01.115
24. C. Hu, J. Guo, J. Wen, Y. Peng, Preparation and electrochemical performance of nano-Co3O4 anode materials from spent li-ion batteries for lithium-ion batteries. J. Mater. Sci. Technol. **29**, 215–220 (2013). https://doi.org/10.1016/J.JMST.2013.01.009
25. C.K. Lee, K.I. Rhee, Preparation of LiCoO2from spent lithium-ion batteries. J. Power Sources **109**, 17–21 (2002). https://doi.org/10.1016/S0378-7753(02)00037-X9
26. H. Nie, L. Xu, D. Song, J. Song, X. Shi, X. Wang, L. Zhang, Z. Yuan, LiCoO2: Recycling from spent batteries and regeneration with solid state synthesis. Green Chem. **17**, 1276–1280 (2015). https://doi.org/10.1039/c4gc01951b

27. M. Lu, H. Zhang, B. Wang, X. Zheng, C. Dai, The re-synthesis of LiCoO 2 from spent lithium ion batteries separated by vacuum-assisted heat-treating method. (2013). www.electrochemsci.org Accessed 18 Jan 2019
28. D.-S. Kim, J.-S. Sohn, C.-K. Lee, J.-H. Lee, K.-S. Han, Y.-I. Lee, Simultaneous separation and renovation of lithium cobalt oxide from the cathode of spent lithium ion rechargeable batteries. J. Power Sources **132**, 145–149 (2004). https://doi.org/10.1016/J.JPOWSOUR.2003.09.046
29. J. Nan, D. Han, X. Zuo, Recovery of metal values from spent lithium-ion batteries with chemical deposition and solvent extraction. J. Power Sources **152**, 278–284 (2005). https://doi.org/10.1016/J.JPOWSOUR.2005.03.134
30. L. Li, R.J. Chen, X.X. Zhang, F. Wu, J. Ge, M. Xie, Preparation and electrochemical properties of re-synthesized LiCoO2from spent lithium-ion batteries. Chinese Sci. Bull. **57**, 4188–4194 (2012). https://doi.org/10.1007/s11434-012-5200-5
31. D. Song, X. Wang, E. Zhou, P. Hou, F. Guo, L. Zhang, Recovery and heat treatment of the Li(Ni1/3Co1/3Mn1/3)O2cathode scrap material for lithium ion battery. J. Power Sources **232**, 348–352 (2013). https://doi.org/10.1016/j.jpowsour.2012.10.072
32. T. Masayasu, Y. Dai, S. Kazuhiko, M. Ryuji, H. Hirotoshi, A. Hidehito, Y. Tatsuya, K. Masato, A. Kazuya, A. Masatoshi, I. Yasuhiro, I. Takayoshi, K. Koji, N. Tatsuya, N. Takeshi, O. Kenichi, O. Takashi, S. Yoshifumi, S. Kaoru, S. Takena, S. Junichi, S. Kentaro, T. Itaru, T. Kentaro, Y. Hideshi, Y. Noboru, H. Tatsumi, Current status of a new polarized neutron reflectometer at the intense pulsed neutron source of the materials and life Science experimental facility (MLF) of J-PARC. Chinese J. Phys. **50**, 161–170 (2012). https://doi.org/10.5297/ser.1201.002
33. M.J. Ganter, B.J. Landi, C.W. Babbitt, A. Anctil, G. Gaustad, Cathode refunctionalization as a lithium ion battery recycling alternative. J. Power Sources **256**, 274–280 (2014). https://doi.org/10.1016/j.jpowsour.2014.01.078
34. Y. Weng, S. Xu, G. Huang, C. Jiang, Synthesis and performance of $Li[(Ni_{1/3}Co_{1/3}Mn_{1/3})_{1-x}Mg_x]O_2$ prepared from spent lithium ion batteries. J. Hazard. Mater. **246–247**, 163–172 (2013). https://doi.org/10.1016/j.jhazmat.2012.12.028
35. S.-K. Kim, D.-H. Yang, J.-S. Sohn, Y.-C. Jung, Resynthesis of $LiCo_{1-x}Mn_xO_2$ as a cathode material for lithium secondary batteries. Met. Mater. Int. **18**, 321–326 (2012). https://doi.org/10.1007/s12540-012-2016-4
36. L. Yao, H. Yao, G. Xi, Y. Feng, Recycling and synthesis of LiNi1/3Co1/3Mn1/3O2 from waste lithium ion batteries using d, l-malic acid. RSC Adv. **6**, 14947–14954 (2016). https://doi.org/10.1039/c5ra25079j
37. J. Senćanski, D. Bajuk-Bogdanović, D. Majstorović, E. Tchernychova, J. Papan, M. Vujković, The synthesis of $Li(Co_{[sbnd]}Mn_{[sbnd]}Ni)O_2$ cathode material from spent-Li ion batteries and the proof of its functionality in aqueous lithium and sodium electrolytic solutions. J. Power Sources **342**, 690–703 (2017). https://doi.org/10.1016/j.jpowsour.2016.12.115
38. D. Bian, Y. Sun, S. Li, Y. Tian, Z. Yang, X. Fan, W. Zhang, A novel process to recycle spent LiFePO4 for synthesizing LiFePO4/C hierarchical microflowers. Electrochim. Acta **190**, 134–140 (2016). https://doi.org/10.1016/J.ELECTACTA.2015.12.114
39. H. Zou, E. Gratz, D. Apelian, Y. Wang, A novel method to recycle mixed cathode materials for lithium ion batteries. Green Chem. **15**, 1183 (2013). https://doi.org/10.1039/c3gc40182k
40. P.H. Duvigneaud, T. Segato, Synthesis and characterisation of LiNi1-x-yCoxAlyO2 cathodes for lithium-ion batteries by the PVA precursor method. J. Eur. Ceram. Soc. **24**, 1375–1380 (2004). https://doi.org/10.1016/S0955-2219(03)00565-X
41. H. Cao, B. Xia, N. Xu, C. Zhang, Structural and electrochemical characteristics of Co and Al co-doped lithium nickelate cathode materials for lithium-ion batteries. J. Alloys Compd. **376**, 282–286 (2004). https://doi.org/10.1016/j.jallcom.2004.01.008
42. X. Yao, J. Kong, X. Tang, D. Zhou, C. Zhao, R. Zhou, X. Lu, Facile synthesis of porous CoFe2O4 nanosheets for lithium-ion battery anodes with enhanced rate capability and cycling stability. RSC Adv. **4**, 27488–27492 (2014). https://doi.org/10.1039/c4ra02835j
43. O. Caltun, I. Dumitru, M. Feder, N. Lupu, H. Chiriac, Substituted cobalt ferrites for sensors applications. J. Magn. Magn. Mater. **320**, e869–e873 (2008). https://doi.org/10.1016/j.jmmm.2008.04.067

44. D.C. Jiles, C.C.H. Lo, The role of new materials in the development of magnetic sensors and actuators. Sens. Actuators A Phys. 3–7 (2003). https://doi.org/10.1016/s0924-4247(03)00255-3
45. L. Yao, Y. Xi, G. Xi, Y. Feng, Synthesis of cobalt ferrite with enhanced magnetostriction properties by the sol-gel-hydrothermal route using spent Li-ion battery. J. Alloys Compd. **680**, 73–79 (2016). https://doi.org/10.1016/j.jallcom.2016.04.092
46. L. Yang, G. Xi, T. Lou, X. Wang, J. Wang, Y. He, Preparation and magnetic performance of Co0.8Fe2.2O4by a sol-gel method using cathode materials of spent Li-ion batteries. Ceram. Int. **42**, 1897–1902 (2016). https://doi.org/10.1016/j.ceramint.2015.09.159
47. S. Ameer, I.H. Gul, M. Mujahid, Ultra low permittivity/loss CoFe2O4 and CoFe2O4–rGO nanohybrids by novel 1-hexanol assisted solvothermal process. J. Alloys Compd. **642**, 78–82 (2015). https://doi.org/10.1016/J.JALLCOM.2015.04.101
48. S. Zhang, Q. Jiao, J. Hu, J. Li, Y. Zhao, H. Li, Q. Wu, Vapor diffusion synthesis of rugby-shaped CoFe2O4/graphene composites as absorbing materials. J. Alloys Compd. **630**, 195–201 (2015). https://doi.org/10.1016/j.jallcom.2015.01.037
49. Y. Yao, Z. Yang, D. Zhang, W. Peng, H. Sun, S. Wang, Magnetic CoFe 2O 4-graphene hybrids: facile synthesis, characterization, and catalytic properties. Ind. Eng. Chem. Res. Am. Chem. Soc. 6044–6051 (2012). https://doi.org/10.1021/ie300271p
50. F. Huixia, C. Baiyi, Z. Deyi, Z. Jianqiang, T. Lin, Preparation and characterization of the cobalt ferrite nano-particles by reverse coprecipitation. J. Magn. Magn. Mater. **356**, 68–72 (2014). https://doi.org/10.1016/j.jmmm.2013.12.033
51. M. Houshiar, F. Zebhi, Z.J. Razi, A. Alidoust, Z. Askari, Synthesis of cobalt ferrite ($CoFe_2O_4$) nanoparticles using combustion, coprecipitation, and precipitation methods: A comparison study of size, structural, and magnetic properties. J. Magn. Magn. Mater. **371**, 43–48 (2014). https://doi.org/10.1016/j.jmmm.2014.06.059
52. R. Subbaraman, D. Tripkovic, K.-C. Chang, D. Strmcnik, A.P. Paulikas, P. Hirunsit, M. Chan, J. Greeley, V. Stamenkovic, N.M. Markovic, Trends in activity for the water electrolyser reactions on 3d M(Ni Co, Fe, Mn) hydr(oxy)oxide catalysts. Nat. Mater. **11**, 550–557 (2012). https://doi.org/10.1038/nmat3313
53. S. Anantharaj, S.R. Ede, K. Sakthikumar, K. Karthick, S. Mishra, S. Kundu, Recent trends and perspectives in electrochemical water splitting with an emphasis on sulfide, Selenide, and Phosphide Catalysts of Fe Co, and Ni: A Review. ACS Catal. **6**, 8069–8097 (2016). https://doi.org/10.1021/acscatal.6b02479
54. S. Anantharaj, P.N. Reddy, S. Kundu, Core-oxidized amorphous cobalt phosphide nanostructures: an advanced and highly efficient oxygen evolution catalyst. Inorg. Chem. **56**, 1742–1756 (2017). https://doi.org/10.1021/acs.inorgchem.6b02929
55. J. Wang, W. Cui, Q. Liu, Z. Xing, A.M. Asiri, X. Sun, Recent progress in cobalt-based heterogeneous catalysts for electrochemical water splitting. Adv. Mater. **28**, 215–230 (2016). https://doi.org/10.1002/adma.201502696
56. S. Natarajan, S. Anantharaj, R.J. Tayade, H.C. Bajaj, S. Kundu, Recovered spinel MnCo2O4 from spent lithium-ion batteries for enhanced electrocatalytic oxygen evolution in alkaline medium. Dalt. Trans. **46**, 14382–14392 (2017). https://doi.org/10.1039/c7dt02613g
57. I.L. Santana, T.F.M. Moreira, M.F.F. Lelis, M.B.J.G. Freitas, Photocatalytic properties of Co3O4/LiCoO2recycled from spent lithium-ion batteries using citric acid as leaching agent. Mater. Chem. Phys. **190**, 38–44 (2017). https://doi.org/10.1016/j.matchemphys.2017.01.003
58. B. Andriyevsky, K. Doll, T. Jacob, Electronic and transport properties of LiCoO2. Phys. Chem. Chem. Phys. **16**, 23412–23420 (2014). https://doi.org/10.1039/c4cp03052d
59. T. Zhao, Y. Yao, M. Wang, R. Chen, Y. Yu, F. Wu, C. Zhang, Preparation of MnO2-modified graphite sorbents from spent li-ion batteries for the treatment of water contaminated by lead, cadmium, and silver. ACS Appl. Mater. Interfaces. **9**, 25369–25376 (2017). https://doi.org/10.1021/acsami.7b07882

Bio-hydrometallurgical Methods For Recycling Spent Lithium-Ion Batteries

Nazanin Bahaloo-Horeh, Farzane Vakilchap and Seyyed Mohammad Mousavi

Abstract A mature technology of lithium-ion batteries (LIBs) is applied in various electronic devices. The wide application of LIBs has brought large quantities of spent batteries, which has become a global problem. Owing to unfavorable effects of spent LIBs on the economic and environmental aspects, much effort has been made in many countries to manage and recycle the waste batteries. Owing to several restrictions in conventional recycling methods, the use of microorganisms has attracted increasing attention. The bio-hydrometallurgical approaches realize the win-win situation of environmental and economic benefits. In this chapter, the information available on the basic principles and recent developments of the bioleaching of metals from LIBs are reviewed in detail. Additionally, this chapter gives an overview of the previous studies performed in this field. Furthermore, the challenges, limitations, and potential solutions for applying more efficient bioleaching approach for recovery of metals from LIBs are highlighted.

Keywords Lithium-ion batteries · Recycling processes · Bio-hydrometallurgy · Bioleaching · Microorganism

1 Introduction

Favorable properties of lithium-ion batteries (LIBs) including a high energy density, low memory effect, good cycle life, high cell voltage, low self-discharge, wide temperature domain of use, long storage life, safety, and lightweight have gradually led to use of them in the global rechargeable battery market as an alternative of the Ni–MH and Ni–Cd batteries [1–4]. In recent years, production and consumption of batteries in large quantity have increased dramatically due to reduced cost and ubiquitous application in industry and communication technology [5, 6]. LIBs have

N. Bahaloo-Horeh and F. Vakilchap—Authors have the same contribution.

N. Bahaloo-Horeh · F. Vakilchap · S. M. Mousavi (✉)
Biotechnology Group, Chemical Engineering Department, Tarbiat Modares University, Tehran, Iran
e-mail: mousavi_m@modares.ac.ir

L. An (ed.), *Recycling of Spent Lithium-Ion Batteries*,
https://doi.org/10.1007/978-3-030-31834-5_7

been applied in portable electronic devices including sensors, laptops, digital cameras, mobile telephones, notebooks, consumer electronics (CE), and electric vehicles (EV) [2]. Over the past decade, the global rate of LIBs production has increased exponentially from 2.04 billion in 2007 to 4.49 $\times$ 10^9 units in 2011 and 7 billion in 2015 [7].

Increase in the production and consumption of LIBs in recent years have extensively augmented the amount of spent LIBs [3]. Production of spent LIBs is increasing worldwide especially with 10% of market share in Korea [8]. Specifically, it is estimated that the weight of spent LIBs will exceed 5 $\times$ 10^5 metric tons and 25 billion units, respectively, in the year 2020 in China [9].

2 Spent LIBs Management

Strict global regulations have been legislating for hazardous wastes on the manufacturing, utilization, amassment, recycling, and disposal of wastes [10]. Spent LIBs are a considerable stream of waste which requires sufficient management due to proliferating utilization of LIBs and growing number of waste LIBs [11]. The Environmental Protection Agency (EPA) take into account all waste batteries as hazardous materials [9]. Because of the enormous amount of waste batteries and both viewpoint of economic and environmental revenue, the technique of recycling is the most effective strategy for waste management [2]. In 2006, the European Union (EU) announced the Directive 2006/66/EC of the European Parliament and the Council on batteries and accumulators. The EU directive obligates all member states to enhance recovery rate of the entire EU's waste batteries from at least 25% in 2012 to 45% in 2016, and at least 50% of LIBs should be recycled. Moreover, WEEE Directive (2012/19/EU) stated that by 2018, 75% of waste mobile phones (containing spent LIBs) must be recycled and reused [12]. In the following, major reasons for spent LIBs recycling are described.

2.1 Environmental Aspect

Owing to presence of hazardous heavy metals (e.g. cathodic materials such as $LiCoO_2$, $LiMn_2O_4$, $LiNiO_2$) and toxic-flammable ingredients (e.g. electrolytes including $LiBF_4$, $Li(SO_2CF_3)_2$, $LiPF_6$, or $LiCF_3SO_3$) [7], disposal of spent LIBs in an inadequate way constituted a potential threat to the human health and ecosystem [13]. Heavy metals and other compounds may cause dangerous contamination in incinerator ash and emissions, compost, and landfill leachate [9] due to heavy metals permanence in the environment and being carcinogenic and toxic even at low concentrations [14].

Most waste LIBs as domestic waste is disposed of in landfills [3]. The leachable metals including Co, Ni, and Li are enclosed to the battery cathode, which is protected by the battery outer casing. Nevertheless, if the outer casing of the battery is broken during the process of compaction or becomes chemically degraded, the components inside of the battery can be exposed to landfill leachate [15]. The heavy metals and organic electrolytes of LIBs can seep and leach gradually into the soil, surface, and groundwater [16]. Also, landfill leachate can transport these pollutants to places outside of the landfill and affect the ecosystem, agriculture, and human health [15]. Owing to metals reduction-oxidation characteristics, complex formation tendencies, and intrinsic solubility at different pH, the metals mobility from LIBs in the aqueous system is a great concern [16]. Moreover, the licensed landfills [17] and landfill capacity spaces for their disposal are confined [3]. Furthermore, spent LIBs incineration with other wastes result in releasing toxic gases into the atmosphere [18].

2.2 Economical Aspect

LIBs can act as a secondary rich and cheap repository of valuable metals such as Co, Li, Mn, Ni, and Al and at times are a richer repository of minerals than primary mineral reserves [17]. On the other hand, the metals contained in LIBs depict a non-negligible resource of valuable metals; thus, the preservation of natural reserves is achieved by waste LIBs recycling [19]. For example, Co is a comparatively expensive metal [8] owing to low abundance in nature [7] and Li has vital significance in plenty of industrial fields [8].

LIBs production has led to the loss of valuable metals. For example, the spent LIBs with cathodic material of $LiCoO_2$ include up to 7% Li and 20% Co [7]. Manufacturing of LIBs dedicated 25% of the worldwide demand for Co. However, recycled metals can be employed as raw materials in different application such as battery production. Also, the required energy for recycling waste LIBs is much less than the energy required for producing LIBs [20].

It was shown that 51% of natural reserves will be preserved if Co and Ni of waste LIBs are recycled instead of mining from primary reserves [21]. Furthermore, owing to a shortage of natural reserves supply and the rising demand, many metals are considered as the critical metals [22]. Table 1 demonstrates the availability of metals contained in the LIBs. Albeit metals including Ni, Cu, and Co are not rare, their natural resources are falling to notably low levels [23]. This phenomenon forces the recycling of waste LIBs [11].

Table 1 The availability of LIBs metals

Element	Li	Al	Co	Ni	Mn	Cu
Availability (years)	>70	65	59	30	29	25

On the other hand, because the environmental global regulations, especially in terms of toxic wastes disposal, become more strict, the charges for verifying environmental protection will continuously rise. Therefore, recycling of spent LIBs is becoming urgent [4].

3 Recycling Methods of Spent LIBs

The EU Battery Directive 2006/66/EC announced the waste batteries management and encourages all member states to create new recycling technologies and develop studies into environmentally friendly and affordable methods [20].

Finding an efficient, safe, eco-friendly and economical recycling technologies is imperative as well as development and optimizing the current recycling programs [3]. Until now, there isn't an effective recycling technology for waste LIBs [9].

General technologies for recycling spent batteries are divided into three categories: conventional processes of pyrometallurgy and hydrometallurgy and a new method of biohydrometallurgy [22].

The pyrometallurgical methods consist of the thermal treatment that causes chemical and physical transformations in the materials in order to recover valuable metals [24]. The pyrometallurgical methods include the processes of volatilization or melting of metals which need high temperatures [22]. Although elevated temperature supports metals leachability, it increases total process costs [25].

In hydrometallurgical methods, strong acids including HCl, HNO_3, and H_2SO_4 along with oxidizing agent such as H_2O_2 leach the LIBs metals [2]. There are many possibilities for changing the conditions of this process to achieve the desired results [25].

Nowadays, bio-hydrometallurgical methods as the application of naturally-occurring microorganisms are gradually replacing the traditional methods [26]. For achieving environmental and metal sustainability, it's necessary to develop green processes for wastes LIBs recycling [20].

In the following, the comparison between these processes and their advantages and disadvantages are described.

3.1 Comparison of Recycling Methods

Although the hydrometallurgical and pyrometallurgical methods are extensively applied for metals extraction from waste batteries, they encompass inherent constraints [27].

Pyrometallurgical processes are fast due to not importance of scrap physical form as that required in chemical treatments [24]. Although the pyrometallurgical method is efficient, fast, relatively simple, and does not need battery dismantling, it generates polluting emissions of dust and hazardous gases such as toxic furans, dioxins, and

SO_2 gas [27]; thus, it needs dust collecting/gas cleaning systems which results in additional investment [22].

Also, since the temperature of the pyrometallurgical processes is often higher than 800–1000 °C, they are very energy intensive [13]. Furthermore, pyrometallurgical operations are extremely harsh, not easy to control, involve high costs, their equipment requirement is high, and they are not versatile and flexible [17, 22]. It must be considered that in pyrometallurgical recycling processes of waste LIBs, the recovery of Li cannot be achieved; which is the main drawback of the pyrometallurgical process [3].

In comparison with pyrometallurgical processes, hydrometallurgical ones are more acceptable because of inherent advantages including elimination of toxic gas emissions as no particles produced, consuming less energy, be more economical, having relatively simpler operation, low cost requirements, a high purity of the final metal product, and possible recovery of leachants [22, 24, 28].

However, the hydrometallurgical process needs pretreatment of the waste LIBs [6]. It is time-consuming, dangerous and not eco-friendly owing to use substantial amounts of strong acids and chemicals reagents [28] which leads to the generation of hazardous secondary pollution and high-security risk [6]. Therefore, this process needs subsequent downstream processing prior to environmental discharge [27].

In comparison with pyrometallurgical and hydrometallurgical methods, bio-hydrometallurgical processes are interesting alternative methods [24] because bio-hydrometallurgical processes are simple, have higher efficiency, high safety, lower capital costs (only one-third to one-half of traditional processes), easier management, lower energy consumption, mild reaction conditions (operation at atmospheric pressure and room temperature), negligible environmental impact, need few industrial requirements along with not requiring skilled workers [9, 20, 29]. Furthermore, biological processes could be more selective toward metals due to the sensitivity of microorganisms toward high concentrations of various metal ions leading to selective metal recovery. In addition, microbiological processes and principles have the potential to promote owing to readily adaptation of the microorganisms to toxic metal ions and conditions [30]. Another advantage of bioleaching is that the necessary leaching reagents for solubilization of metal are biologically generated and continuous delivery of reagents is not needed [31]. Lastly, the bio-hydrometallurgical technologies are performed in a closed loop which produces minimum effluents [32].

Nonetheless, slow kinetics of bio-hydrometallurgical processes is their great disadvantage. Longer operation time is needed to obtain reasonable yields compared to traditional processes [33]. Another disadvantage of the bio-hydrometallurgical method is related to the process control measures (temperature, oxygen, pH, and nutrients) [16]. Another concern is about optimum particle size maintenance in the bioreactor. The activity of cells can negatively be influenced by too fine particles (<25 μm) because of an increased particle-particle collision which resulted in attrition. Furthermore, bioleaching of industrial wastes (e.g. waste LIBs) is mostly performed with waste load (ratio of solid to liquid). The efficiency of the bioleaching process at higher pulp density decreases owing to metal toxicity or alkalinity of wastes to microorganisms [33]. The merit of bio-hydrometallurgical processes in achieving the purposes of "low environmental degradation and resource reclamation" can offset these disadvantages [16].

In recent years, for improving bioleaching efficiency, researchers looked into different aspects including the addition of surfactants, accession of metal ions, improvement of bacterial culture, and breeding efficient bacteria used in bioleaching [5]. It is suggested that the bioleaching efficiency can also be improved by some pretreatment methods such as microwave and ultrasound [34]. However, more research is needed to enhance recovery of the metals and recycling of spent LIBs in these areas [9].

4 Bioleaching (Microbial-Based Recovery)

Bioleaching or microbial-based recovery has been employed industrially for several decades without operators knowing the microorganism's roles in solubilization of metal until 1961 when the bacteria of *Acidithiobacillus ferrooxidans* (which oxidizes Fe^{2+} to Fe^{3+}) was discovered in the leachates. Since 1980, the bioleaching method has been efficiently applied to industrial mining plants in various locations in order to get a better understanding of used microbes [34, 35]. This progress proved that the bioleaching process is a practical method in the case of primary reserves processing, especially lower grades. While today, the use of bioleaching is more about the secondary materials that often show dissimilarities with the bioleaching of primary ores because in the secondary materials most metals are in the zero-valent metallic form in the combination with other metals in solid matrices [30].

Generally, bioleaching is a process as the dissolution of metals involved in mineral sources or secondary wastes by the use of certain microorganisms for the extraction of elements from them [36]. The metabolites excreted by microbial activity or the enzymatic oxidation-reduction on the solid matrix can leach out the metals from toxic waste materials [7, 9, 30, 37]. Microbial metabolites derived from organisms assist in the bioleaching through proton-promoted mechanisms and/or ligand-promoted [38].

Bioleaching experiments are usually performed through three approaches consisting of one-step, two-step, and spent-medium [9, 29]. The one-step approach is a classical method through exposure to biomass which is conducted via inoculation of microbial pure culture or spore suspension to the culture medium containing solid waste (for example, waste LIBs) [7, 9]. In one-step fungal leaching, the operating and capital costs are reduced because fermentation and leaching processes are done in one stage [20].

In the two-step approach, organism (bacteria or fungi) grows in the culture medium until the logarithmic growth phase, and after that, the solid waste is added to the medium [7, 29]. It should be noted that in the case of fungi, a sudden reduction in pH (caused by organic acid secretion) is a symptom of entering the logarithmic phase [3]. In two-step fungal leaching, the germination would limit the inhibition effect of the solid powder on fungal growth, production of organic acids and correspondingly on metal extraction. Also, the organic acids excretion prior to addition of solid powder to the medium decreases the heavy metal ions toxicity through complexation [39].

It's worth noting that the toxicity limits the optimal applied pulp density and has not a favorable influence on direct leaching [40]. It was proposed that a two-step approach for an industrial application can be appropriate due to enhanced bioleaching efficiency caused by acidic solution generated by microorganism before the addition of solid powder to the medium [32].

In the spent-medium approach, a pure culture of microbes is performed at first until the stationary phase of microbe growth. Then, the suspension of mycelia in a fungal liquid medium or bacterial cells are centrifuged or filtered in order to obtain a cell-free medium. At last, leaching can be performed by solid waste addition to the filtrate (cell-free medium) including metabolites such as mineral or organic bio-acids [3, 7, 29].

Several comparative studies have been demonstrated in the case of these three approaches, and in most cases, the highest bioleaching performance was achieved in the spent-medium bioleaching approach. There are advantages in spent-medium bioleaching; for example, there is no direct contact between the solid particles and microbes; thus, each of the processes (chemical and biological) can be optimized independently. For example, the quantity and quality of bio-produced acid can be optimized in the fermentation stage for maximizing productivity. In the leaching stage, there is no limitation of applying the high pulp densities due to not existence of microbial cells and their cell wall damage by the high toxicity of batteries [7]. Since, in spent-medium bioleaching, the fermentation of microbes performs in one stage and the bio-generated acids are used in other stages, the capital costs would enhance owing to extra bioreactor tanks requirements [20].

Bahaloo-Horeh et al. [3] compared the different approach of bioleaching (one-step, two-step, and spent-medium bioleaching) for metals recovery from waste LIBs using *Aspergillus niger* fungi. Among different approach, the highest recovery of metals was achieved in spent-medium bioleaching.

Biswal et al. [7] applied *Aspergillus niger* MM1/SG1 strains and *Acidithiobacillus thiooxidans* 80,191 in order to recover lithium and cobalt from waste LIBs powder under one-step and spent-medium approaches. The metal removal in fungal leaching via spent-medium approach was comparable or even better than bioleaching by bacteria. The comparison between fungal and bacterial leaching in the spent-medium approach demonstrated that 100% of lithium and 80–82% of cobalt recovered by *Aspergillus niger* strains while 66% of lithium and 23% of cobalt recovered by *Acidithiobacillus thiooxidans*.

Bahaloo-Horeh and Mousavi [9] applied spent-medium bioleaching for metals recovery from LIBs. In the first stage, optimization of organic acids secreted by *Aspergillus niger* was done by finding the desired fungal cultivation conditions. Organic acids including gluconic, citric, and malic acid were produced at maximum concentrations of 8433.76, 26,478, and 1832.53 ppm, respectively. In the next stage, they used the bio-produced organic acids to investigate the recovery of metals at different pulp densities of waste LIBs. They found out the recovery of 77% manganese, 100% copper, 75% aluminum, and 100% lithium at pulp density of 2% (w/v) but 54% nickel and 64% cobalt at pulp density of 1% (w/v) [9].

4.1 Adaptation: A Tool for Improving One-Step and Two-Step Bioleaching

Heavy metals toxicity has a significant effect on the microbes growth and leaching function during the one-step and two-step bioleaching. The metabolic pathways have an effective role in the generation of biogenic acids that can be inhibited by toxicity characteristic of metals during the leaching process. Heavy metals act as a protoplasmic poison and also can induce the denaturation of macromolecules such as proteins and nucleic acids. However, microorganisms can survive through adaptation or mutation when toxic heavy metals exist in the bioleaching process [20]. The progress of a microorganism's tolerance to the elevated concentrations of metals that is called an adaptation step can be created through microbial pre-grow along with the gradual addition of solid waste powder into solution [38]. The biochemical pathways play a crucial role in the microbes adaptation to toxicity derived from the existence of heavy metals at high concentrations [20]. Furthermore, mutation breeding is an effective method to obtain microorganisms with high-performance and high-tolerance to the toxicity of wastes [34].

The adaptation process before bioleaching can increase the leaching efficiency compared to bioleaching by using unadapted microorganism [20]. Due to the toxic nature of LIBs, most of the previous studies applied adaptation before the bioleaching process [4, 5, 8, 41]. Bahaloo-Horeh et al. [20] reported that by adaptation of *Aspergillus niger*, the organic acids secretion and bioleaching performance were improved. Moreover, the rate of acid secretion was increased; in addition to the reduction in a period of time required for entrance to the logarithmic growth phase. The adapted fungi produced mainly gluconic acid in the presence of waste LIBs and leached 94% copper, 100% lithium, 62% aluminum, 72% manganese, 38% cobalt, and 45% nickel from 1% (w/v) waste LIBs [20].

Hartono et al. [11] isolated local bacterial strains from mud and soil sample with the ability to extract Li from spent LIBs. The isolated bacteria accumulated to lithium and cobalt using LiCl and $CoCl_2$ solution, respectively. Co-adapted and Li-adapted bacteria were used in bacterial leaching experiments. It was shown that bacterial adaptation with LiCl solution can enhance the lithium recovery; maximum lithium recovery was 62.83% after 15 days.

5 Microbes in Bioleaching

Microorganisms as miniaturized chemical bioreactors enable to carry out any chemical reaction as a catalyst under controlled pressure and temperature [19]. Different microorganisms should be tested and prudently chosen for achieving the best bioleaching yield [38]. Three groups of microbes which commonly use in bioleaching approach are autotrophic bacteria, heterotrophic fungi and bacteria [39]. In the following section, these groups are introduced.

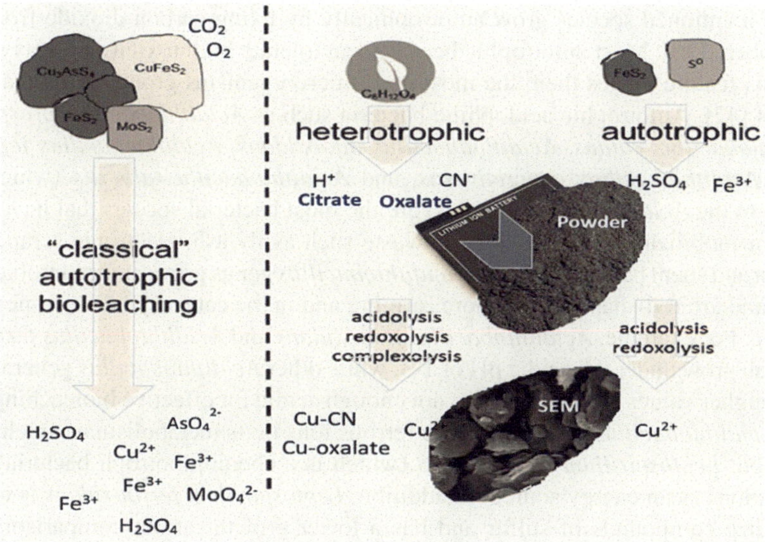

Fig. 1 An overview of autotrophic and heterotrophic bioleaching of secondary resources and conventional autotrophic bioleaching of primary resources [42]

It's worth noting that in general, bioleaching mechanisms of primary ores are different from mechanisms of secondary resources due to fundamentally different chemistry of metals presented in primary and secondary resources. The waste batteries are non-sulfidic material and consist of elemental metals, metal oxides and hydroxides [28]. Figure 1 demonstrates an overview of autotrophic and heterotrophic bioleaching of secondary resources and conventional autotrophic bioleaching of primary resources.

5.1 Autotrophic Bacteria

The autotrophic microbes can obtain their energy by oxidation of the low-valence sulfur/iron as an electron donor in a manner of adhesion between cells and solid waste [43]. That means these species do not require organic carbon as a carbon source for their growth [34]. In the metabolism of the chemolithoautotrophs, the terminal electron acceptor (or the oxidant) is molecular O_2 that is required for the respiratory reaction according to following half equation [36]:

$$O_2 + 4H^+ + 4e \rightarrow 2H_2O \tag{1}$$

The mentioned species grow autotrophically by fixing carbon dioxide from the atmosphere [36]. Most autotrophic bacteria can tolerate high toxicity of heavy metals. This feature makes them the most used microorganisms group for bioleaching process [42]. Autotrophic acidophilic bacteria such as *Acidithiobacillus prosperus*, *Acidithiobacillus caldus*, *Acidithiobacillus thiooxidans*, *Acidithiobacillus ferrooxidans, Acidithiobacillus concretivorus*, and *Acidithiobacillus albertis* (which are belong to the *Acidithiobacillus* genus) are the most bacterial species that have been used for mobilization of metals from waste such as fly ash, electronic scrap, used catalyst and spent batteries [7]. The *Acidithiobacillus* genus grows under aerobic condition and are rod-shaped, non-spore forming and in the category of Gram-negative bacteria. For example, *Acidithiobacillus thiooxidans* and *Acidithiobacillus ferrooxidans* can grow under the acidic pH of 1-3, while other *Acidithiobacillus* genera grow under higher values of pH which are not enough acidic for effective bioleaching process. *Acidithiobacillus caldus* not uses ferrous ions in its metabolism as an electron donor but *Leptospirillum ferrooxidans* (which is a chemolithotroph bacteria) uses ferrous ions as an energy source. In addition, *Leptospirillum ferrooxidans* is unable to oxidize compounds of sulfur and has a lower growth rate in comparison with *Acidithiobacillus ferrooxidans* which can be catalyzed by adding Zn^{2+} ions [34].

Indeed, acidophilic sulfur-oxidizing bacteria (SOB) such as *Acidithiobacillus thiooxidans* and iron-oxidizing bacteria (IOB) such as *Acidithiobacillus ferrooxidans* are the most widely used bacteria in the bioleaching process [44]. *Acidithiobacillus ferrooxidans* as a powerful oxidizing agent acts for oxidizing Fe^{2+} to Fe^{3+} ions and *Acidithiobacillus thiooxidans* oxidizes sulfur compounds such as elemental sulfur, sulfides, and thiosulfate (as energy sources) to sulfate (as a final oxidation product). Through these actions, the pH value becomes as low as 1 in the leachate and creates suitable acidic conditions for their growth. Both of these acidophilic bacteria use CO_2 and oxygen as a carbon source and the terminal electron acceptor, respectively [28, 29, 34].

In another classification, these bacteria can be divided based on the feasible temperature range for their growth. Most industrial microorganisms are in the group of mesophilic bacteria such as *Acidithiobacillus* and *Leptospirillum* genera that the preferred temperature for their growth is about 25–35 °C. Another group is a thermophilic bacteria that can tolerate the high temperatures. These microbes are desired for the process of bioleaching owing to their high microbial activities in exothermic biooxidation in stirred bioreactors. The use of these bacteria in the high operating temperatures improves reaction rates and extraction yields of metals from solid wastes [34]. Moderately thermophilic bacteria have ability to grow at a temperature around 50 °C but extreme thermophiles can grow at a temperature above 60 °C which are usually archaea [45]. *Acidianus brierleyi* is one example of archaea which are classified in the category of *sulfolobus* genus; *sulfolobus* species have ability to use the energy sources such as Fe^{2+}, S^0, and sulfides [34].

5.2 Heterotrophic Bacteria and Fungi

Organic materials as energy sources have a significant role in the metabolism of heterotrophic bacteria and fungi. In the growth phase, secretion of metabolites such as organic acids is the result of organic carbon source metabolism [46]. These compounds have ability to solubilize metals by soluble complexes of metal and chelates from a solid matrix. As spent LIBs does not contain metal sulfides, using heterotrophic microorganism for bioleaching of LIBs is a promising approach [42]. Generally, heterotrophic microorganisms are known for their ability to leach materials in the form of oxidic, siliceous or carbonaceous [39]. Furthermore, heterotrophic microorganism could secrete non-acidic complexing agents as a result of the proteins catabolism that could be favorable for alkaline leaching systems [46]. Several fungi such as *Aspergillus niger*, *Penicillium simplicissimum*, and *Penicillium chrysogenum* and bacteria such as *Gluconobacter oxydans*, *Pseudomonas* strains, and *Chromobacterium violaceum* have been used for metals extraction from different wastes [42].

Pseudomonas strains, and *Chromobacterium violaceum*, as well as *Bacillus megaterium*, are belonged to the cyanogenic groups of microorganism. The cyanogenic bacteria produce cyanide in the aqueous medium as a secondary metabolite through decarboxylation of glycine by the enzyme of HCN synthase [47]. Cyanide is produced in a short period of time at the end of the logarithmic growth phase or in the stationary phase [48]. Then, the β-cyanoalanine synthase in some of these bacteria can reduce the cyanide content at the end of the stationary or death phases by conversion cyanide to β-cyanolanin [47].

Cyanide is a mineral compound, poisonous liquid and colorless [49] that its production occurs in two forms including free cyanide anion (CN^-) and the hydrocyanic acid (HCN) in an equilibrium reversible reaction as shown in Eq. (2) [50]. It is estimated that the formation of metal-cyanide complexes is improved at higher pH values due to the pK_a of hydrocyanic acid ($pK_a \sim 9.21$) at 25 °C [51].

$$HCN \leftrightarrow H^+ + CN^- \quad PK_a{}^0(HCN) = 9.21 \pm 0.02\,\text{at}\,25\,°C \tag{2}$$

A specific dissolution mechanism as complexolysis is the only dominant mechanism in the bioleaching by cyanide [52]. The complex of various metal ions (Cu, Ni, Au, Ag, Pt, Pd, Rh, etc.) with cyanide is feasible due to their high chemical stability [53]. However, the interaction between metal ions and bio-cyanide is not well known yet [54].

At the present time, no pilot-scale industrial project of heterotrophic bioleaching process has been performed may be due to the economic unfeasibility of high supply of energy and carbon sources for growth and activity of the heterotrophic microbes [30].

5.3 Comparison of Bacterial and Fungal Leaching

Owing to constant nutrients supply for growth of fungi, handling filamentous fungi in turnover and long processing time was magnified [17]. Also, compared to fungi, bacteria show a higher capability for initial surface metal binding. This indicates that when the rapid binding followed by metal removal is needed, the bacterial system is preferred [36].

However, compared to bacteria, fungal bioleaching has several advantages:

- more ability for tolerating toxic materials (toxicity of elements is the major issue that affects the metals bioleachability [40])
- grow over a broad range of pH values (fungi can grow between pH range of 2–8 [46])
- shorter lag phase
- faster leaching rates
- effective for bioleaching of acid-consuming and alkaline substances
- produce organic acids and formation of metal-complexes [3]
- greater uptake and intracellular accumulation of metals [36].

6 Fungal Leaching

Organic carbon sources are needed for energy supply and the growth of fungi. During this consumption, fungi produce metabolic products [3] including extracellular polymeric substances (EPS), proteins, amino acids, lipids, exopolysaccharides, organic acids or complexing agents like siderophores which can play effective roles in dissolving the metallic portions of waste [55]. Among fungi, *Aspergillus niger* exhibits good potential in the bioleaching process of LIBs due to secretion of abundant chelating agents and organic acids and its ability to grow in highly alkaline medium [24]. Bahaloo-Horeh et al. [3] applied the *Aspergillus niger*, in the first attempt for metals mobilization from spent LIBs. They extracted 70% Mn, 95% Li, 38% Ni, 100% Cu, 65% Al, and 45% Co from waste mobile phone LIBs at 10 (g/L) pulp density through spent-medium bioleaching. The authors showed that higher metals recovery could be achieved by bioleaching compared to the chemical leaching [3]. Biswal et al. [7] also applied spent-medium bioleaching to recover Li and Co from waste LIB using *Aspergillus niger* SG1 and MM1 strains at 0.25% (w/v) pulp density. The findings of this research demonstrate the dissolution of 82% Co and 100% Li. The authors observed no growth of fungus in presence of 0.5% (w/v) waste load.

Fungi have several resistance mechanisms to cope with toxic metals: they have ability to (1) metals immobilization outside their fungal cell with aid of chelating compounds; (2) metals binding into cell wall; (3) metals pumping actively outside the cell; (4) metals binding on particular intracytoplasmic proteins (for instance, glutathione, metallothionein); and (5) metals immobilization into organelles (for instance, into vacuoles). All these properties indicate that fungi are an interesting appliance in metals recovery from waste [19].

Various parameters influence the efficiency of fungi bioleaching which should be considered for the development of the process. These parameters including the microorganisms tolerance to heavy metals, biomass concentration, amount and type of secreted metabolites, physicochemical factors such as composition of media, pH, oxygen supply, temperature, contact time, stirring rate, and material properties such as grain size, pulp density, surface properties, and mineral composition [40]. Bahaloo-Horeh and Mousavi [9] used the organic acids which were excreted by *Aspergillus niger* under optimal conditions to investigate the recovery of metals at different pulp densities of waste LIBs. The results showed that the maximum bioleaching recovery of 75% Al, 100% Li, 77% Mn, and 100% Cu in a pulp density of 20 (g/L), and 54% Ni and 64% Co in a pulp density of 10 (g/L) were achieved [9].

6.1 Fungal Metabolite: Organic Acids

It was shown that among the secreted metabolites, organic acids have a more significant role in metals leaching owing to supplying ligands (organic acid anions) and protons. These organic acids include gluconic, citric, malic, oxalic, fumaric, lactic, pyruvic and succinic acids which are biologically degradable [40]. Also, bioleaching using organic acids is performed under moderately acidic conditions (pH range of 3–5). It's worth noting that chemically-produced organic acids are not a proper choice owing to the low yields along with complicated and expensive processes. That is why the microbiological production of these organic acids is a significant alternative to chemically-produced ones [9].

The concentration, type, and nature of organic acids generated by fungi are affected by the biological, chemical, and physical factors including the of carbon source (its type and concentration), composition of leaching media, pH and buffering capacity of the microbial culture medium, nature of waste materials, and so forth [56]. Bahaloo-Horeh et al. [20] used the response surface methodology to examine the effects of parameters including initial pH, carbon source (sucrose) concentration, and inoculum size on optimization of organic acids secretion by *Aspergillus niger*. Maximum gluconic, malic, and citric acid concentrations of 8433.76, 1832.53, and 26,478 ppm, respectively were obtained under optimal conditions of 116.90 (g/L) carbon source (sucrose) concentration, 3.45% (v/v) inoculum size, and a pH value of 5.44 [9]. It has been revealed by Bahaloo-Horeh et al. [3] that the concentration and type of organic acid generated by *Aspergillus niger* rely on the nature of the metals of LIBs presented in the culture medium. In the absence of LIBs, *Aspergillus niger* secreted citric acid as a major organic acid produced in the culture medium, while in the presence of LIBs, the oxalic acid was secreted as a major organic acid. The presence of metal ions in the culture medium such as copper and manganese may be the reason for lower secretion of citric acid in the presence of LIBs powder [3].

Table 2 The K_a values of some organic acids and reactions of metal-organic acids

Organic acid	Reaction	Pk_a	Equations
Gluconic acid	$C_6H_{12}O_7 \rightarrow C_6H_{11}O_7^- + H^+$	3.86	(3)
	$n[C_6H_{11}O_7^-] + M^{n+} \rightarrow M[C_6H_{11}O_7]_n$		(4)
Oxalic acid	$C_2H_2O_4 \rightarrow C_2HO_4^- + H^+$	1.25	(5)
	$C_2HO_4^- \rightarrow C_2O_4^{2-} + H^+$	4.14	(6)
	$n[C_2HO_4^-] + M^{n+} \rightarrow M[C_2HO_4]_n$		(7)
	$n\left[C_2O_4^{2-}\right] + 2M^{n+} \rightarrow M_2[C_2O_4]_n$		(8)
Malic acid	$C_4H_6O_5 \rightarrow C_4H_5O_5^- + H^+$	3.40	(9)
	$C_4H_5O_5^- \rightarrow C_4H_4O_5^{2-} + H^+$	5.11	(10)
	$n[C_4H_5O_5^-] + M^{n+} \rightarrow M[C_4H_5O_5]_n$		(11)
	$n\left[C_4O_5^{2-}\right] + 2M^{n+} \rightarrow M_2[C_4O_5]_n$		(12)
Citric acid	$C_6H_8O_7 \rightarrow C_6H_7O_7^- + H^+$	3.09	(13)
	$C_6H_7O_7^- \rightarrow C_6H_6O_7^{2-} + H^+$	4.75	(14)
	$C_6H_6O_7^{2-} \rightarrow C_6H_5O_7^{3-} + H^+$	6.40	(15)
	$n[C_6H_7O_7^-] + M^{n+} \rightarrow M[C_6H_7O_7]_n$		(16)
	$n\left[C_6H_6O_7^{2-}\right] + 2M^{n+} \rightarrow M_2[C_6H_6O_7]_n$		(17)
	$n\left[C_6H_5O_7^{3-}\right] + 3M^{n+} \rightarrow M_3[C_6H_5O_7]_n$		(18)

It's also worth noting that higher organic acid concentrations play a more effective role in metals leaching which demonstrated the significance of metals complexation and surface complexation reactions as a metal mobilization mechanism [57].

It can be realized the effect of organic acids from acid dissociation constants (K_a values), which demonstrate the ionization degree. The K_a values of some organic acids and reactions of metal-organic acids are demonstrated in Table 2 [58]. The K_a values are associated with the ionic strength of the solution which is needed for interaction with cations of metals [9].

Organic acid could be monoprotic (for instance, gluconic acid), diprotic (for instance, malic acid and oxalic acid) or triprotic (for instance, citric acid). More carboxyl groups in the structure of organic acid are advantageous for metal mobilization due to the formation of stable ligands. For example, citric acid (with three –COOH groups) is more appropriate to mobilize metals than malic and oxalic acid (with two –COOH groups) [9]. As said before, one of the cathodic materials of LIBs is $LiCoO_2$. As an example, the leaching reactions of cathodic material of $LiCoO_2$ with malic acid are shown in the Eqs. (19) and (20) [59].

$$4LiCoO_2(s) + 12C_4H_6O_5(aq) \rightarrow 4LiC_4H_5O_5(aq) + 4Co(C_4H_6O_5)_2(aq) + 6H_2O(l) + O_2 \quad (19)$$

$$4LiCoO_2(s) + 12C_4H_5O_5^-(aq) + 4Li^+(aq) + 4Co^{2+}(aq.) \rightarrow 4Li_2C_4H_4O_5(aq) + 8CoC_4H_4O_5(aq) + 6H_2O(l) + O_2 \quad (20)$$

Owing to precipitation of metal oxalates, the negative effect of oxalic acid on the mobilization of several metals from waste LIBs was observed. In fact, after fungal production of oxalic acid, the soluble metal ions were immobilized and insoluble oxalates were formed which caused a reduction in the bioavailability [20]. As an example, the reactions between a cathodic material of $LiCoO_2$ and oxalic acid are demonstrated in the Eqs. (21) and (22) [60].

$$2LiCoO_2(s) + 7C_2H_2O_4(aq) \rightarrow 2LiC_2HO_4(aq) + 2Co(C_2HO_4)_2(aq) + 4H_2O(l) + 2CO_2(g) \quad (21)$$

$$2LiCoO_2(s) + 4C_2H_2O_4(aq) \rightarrow Li_2C_2O_4(aq) + 2CoC_2O_4(s) + 4H_2O(l) + 2CO_2(g) \quad (22)$$

Bahaloo-Horeh and Mousavi [9] reported that in bioleaching of metals from mobile phone LIBs, citric acid compared to other secreted organic acids by *Aspergillus niger* (malic, oxalic, and gluconic acid) had a more significant role. Furthermore, the authors reported that lower bioleaching efficiency of nickel and cobalt (54% Ni and 64% Co recovery at 10% (w/v) waste load) resulted due to the formation of insoluble cobalt and nickel oxalate.

However, oxalic acid has a significant role in Al solubilization. It was also reported that citric acid and gluconic acid could effectively leach Cu. As an example, the reaction between citric acid and copper oxide is shown in Eq. (23) [20]:

$$3CuO + 2C_6H_8O_7 \rightarrow Cu_3(C_6H_5O_7)_2 + 3H_2O \quad (23)$$

Bahaloo-Horeh et al. [20] also reported that by adapted *Aspergillus niger* which produce mainly gluconic acid, nearly 94% Cu was recovered from waste LIBs.

6.2 Mechanism of Fungal Leaching

As said before, the mechanism of fungal bioleaching is associated with the production of low molecular weight metabolites such as organic acids which act as bio-lixiviants for metals solubilization from waste matrix through several mechanisms that can take place separately or simultaneously [9]. The possible main mechanisms of fungal bioleaching are based on acidolysis, complexolysis, redoxolysis, biosorption, and bioaccumulation. The first three mechanisms occur via excreted fungal metabolites. If the metal ion is accumulated by fungi from the leached solution, the mechanism of bioaccumulation can be observed [39].

The mechanisms of acidolysis and complexolysis were most important for metals mobilization in leaching by fungi. In the leaching solution, the organic acids deliver protons and anions for mobilization of metals by acidolysis and complexolysis [7].

6.2.1 Acidolysis Mechanism

In the acidolysis mechanism, the mobilization of metal occurs through organic acids secreted by a fungal microorganism [7]. The acid production is a protonic reaction weakening the critical bonds and solubilizing metals [32]. In fact, the protonation of oxygen atoms is taken place which covers the metallic compound's surface. Then, the association of oxygen and protons with water resulted in detaching of metal from the surface [39]. It was proved that low pH is beneficial for acidolysis, with the enhanced release of free cations of metal by protonation [7]. In fact, higher acidity results in weaker chemical bonds in the waste matrix and promote metals dissolution [20].

As an example, Eq. (24) demonstrates the acidolysis reaction which produces nickel ions [3]:

$$NiO + 2H^+ \rightarrow Ni^{2+} + H_2O \tag{24}$$

The mechanism of acidolysis is as same as the mechanism of conventional acid leaching [20].

6.2.2 Complexolysis Mechanism

In complexolysis mechanism, organic acids and other chelating agents such as siderophore detach metals from the waste matrix into soluble metal complexes by chelation. The carboxyl and hydroxyl groups of organic acids cause the formation of complex molecules [9]. The ligand exchange polarizing critical bonds are responsible for the constitution of complexes on metal surfaces, which making possible the metals species detachment from the surface [39]. Moreover, metal ions that are mobilized through the mechanism of acidolysis, are stabilized by the mechanism of complexolysis [3]. It's worth noting that metal ions toxicity can be reduced for microorganism by the stability of metal complexes [39].

Metal complexation depends on several parameters such as pH, concentrations of organic acid anions and metal cations in the solution, and the stability constants of the different complexes [20].

As an example, Eq. (25) demonstrates the complexolysis reaction which produces nickel citrate [3]:

$$Ni^{2+} + C_6H_8O_7 \rightarrow Ni(C_6H_5O_7)^- + 3H^+ \tag{25}$$

It was reported that amino acids are excreted together with organic acids. Both of them provide complexing capacity and protons for metals leaching. Nevertheless, fungal leaching by amino acids has not a highly significant role in metal solubilization due to rare efflux of amino acid among filamentous fungi [39].

6.2.3 Redoxolysis Mechanism

In redoxolysis mechanism, oxidation-reduction reactions are the reason from metals mobilization from the solids matrix. As a result, the mobility of metal is increased based upon the metal type, its oxidation state and its mineral phase [32]. The redoxolysis mechanism has minor importance in bioleaching processes by fungi, but major importance in bioleaching processes by chemolithoautotrophic bacteria. As in the case of fungal leaching of manganese, part of metal mobilization may be owing to the enzymatic reduction of oxidized compounds. Equation (26) demonstrates the redoxolysis reaction which produces manganese ion [39]:

$$MnO_2 + 2e^- + 4H^+ \rightarrow Mn^{2+} + 2H_2O \tag{26}$$

6.2.4 Biosorption Mechanism

Biosorption mechanism is a physicochemical passive process between the bio-components or surfaces (for example, certain biomolecules or biomass) and metal ions in a solution that allows binding the metal ions from solutions [55]. Once the metal biosorbed onto certain biomolecules or biomass, it has to be recovered. Owing to the mycelial lifestyle of filamentous fungi and their very high ratio of surface to volume, they are considered as an attractive biosorbents [19]. The microbial cell wall is highly reactive surfaces and composed of various functional groups (e.g. amine, phosphate, carboxyl, and hydroxyl groups) which can bind metal ions to a lesser or greater extent [32].

At neutral pH conditions, microbe cells have a net negative charge. Due to the functional groups existing within the cell wall, the microbes can lose or gain protons as represented in Eqs. (27)–(29) [36]:

$$R - COOH \rightarrow R - COO^- + H^+ \tag{27}$$

$$R - POH \rightarrow R - PO^- + H^+ \tag{28}$$

$$R - OH \rightarrow R - O^- + H^+ \tag{29}$$

These functional groups are able to attract and bind the ions of metals that have a net positive charge (Eq. (30)). Thus, through the energy independent processes, the metal ion is rapidly bound to the cell surface.

$$R - COO^- + M^{z+} = R - COOM^{(z-1)} \quad (30)$$

These reactions are mostly reversible. The removal of surface-bound metals from the cell wall easily occurs by dilute acid or a chelating agent [36].

It's also worth noting that biosorption can involve both nonliving and living biomass [19]; thus, nutrients are not required in the case of nonliving biomass and the process can be done in highly toxic environments [40]. Besides this, the efficiency of metal biosorption is affected by the type of sorbent, biosorbent dosage, the area of biomass surface, and physicochemical factors including pH, temperature, and other interfering ions concentration [42].

Biosorption advantages are including: (1) energy efficient, (2) easy elements harvest, (3) the possibility of selectively elements absorption at very low concentrations, (4) regeneration of some biosorbents, and (5) mild conditions in terms of temperature and pH [42].

6.2.5 Bioaccumulation Mechanism

Bioaccumulation is a specific type of biosorption in which incorporation of metal ions occurs inside living biomass as a protection way against toxicity of metals. This incorporation is based on active uptake of soluble metals through the cell membrane, but passive uptake of metal ions can also be involved [19]. The bioaccumulation is an energy-dependent process and depends on the active respiration mode of the organism. This uptake could be blocked through respiratory poisons or by anaerobiosis which inhibits energy conservation or aerobic respiration. When the uptake of a metal ion into the microbial cell occurred, the equivalent charge ions are released from the cytoplasm. The cations displaced may be potassium, magnesium, or proton which depends on the type of microorganism [36].

The bioaccumulation phenomena cause the continuous metal solubilization due to disturbing the equilibrium between dissolved ions and solid metal. This matter could be construed as the functioning of mycelium as a sink for metal ions. It has been reported that among the filamentous fungi, *Penicillium* and *Aspergillus* genera have a high ability for metals accumulation from their external environment [39].

7 Bacterial Leaching

7.1 *Mechanism of Bacterial Leaching*

In general, two types of major mechanisms are proposed in bacterial leaching: direct and indirect mechanism (or contact and non-contact bioleaching) [4]. Direct mechanism occurs in one-step bioleaching (simultaneously presence of microorganism and the solid particles in the medium), and two-step process (adding the solid particles in the medium when the microorganisms are in the logarithmic phase of growth). Indirect mechanism occurs in spent-medium bioleaching (adding the solid particles to produced bio-metabolites after a specified time in the absence of microorganism) [39]. Figure 2 shows direct and indirect mechanisms in bacterial bioleaching approach.

7.1.1 Contact Bioleaching Mechanisms (or Direct Bioleaching Mechanisms)

In the direct mechanism, the cells attach to the solid particles (physical contact), but not to the whole solid surface. The cells adsorption to the solid particles occur within some minutes or hours [46]. A two-level interaction plays important roles in this adhesion. The first level is electrostatic force which motivates physical sorption. The leaching environments usually occur in low pH and the positively charged cell envelopes of microbe lead to an electrostatic interaction of the mineral phase. The second level represents the chemical bonds formed between cells and minerals [61]. It

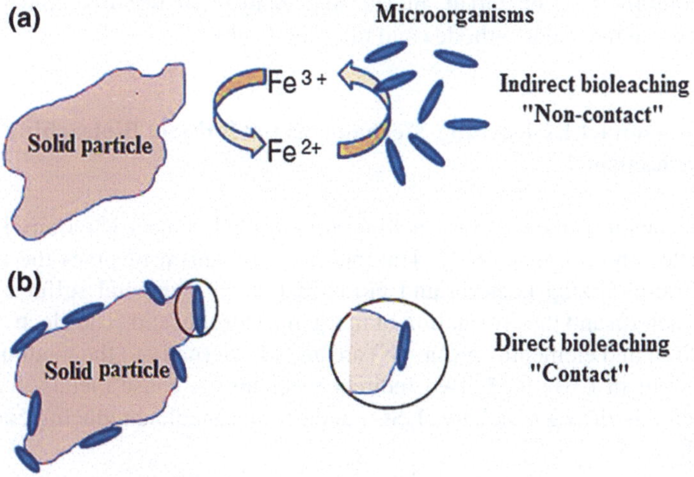

Fig. 2 Direct and indirect mechanisms involved in bioleaching process [46]

was reported that the interaction between cells of *Acidithiobacillus ferrooxidans* and crushed LIBs particles is suspected favorable if the van der Waals force is stronger than the electrostatic force, which would only happen at high ionic strength [42].

After the adhesion of bacteria to solid particles, the microorganisms excrete an exo-polysaccharide layer namely extracellular polymeric substances (EPS) [32]. The EPS is a mixture of an undefined complex of biopolymers containing polysaccharides, lipids, proteins and also humic and nucleic acids that this composition depends on microorganisms type, the age of biofilm and surface properties related to environmental conditions [40]. It was reported that the EPS dosage and components produced by the mixed culture of *Leptospirillum ferriphilum*, *Acidithiobacillus ferrooxidans*, and *Acidithiobacillus thiooxidans* differ from the EPS generated using other types of *Acidithiobacillus* sp. [62, 63].

Through the EPS, a direct bacterial attack including the enzymatic attack and electrochemical interactions occur which leads to metal solubilization [46]. Indeed, in the direct mechanism, the leaching agents generation and the space creation for leaching reactions are taken place and the microorganisms obtain electrons directly from the solid waste and cause the metal solubilization [36]. Furthermore, the bio-oxidation reactions happen most quickly and effectively within the EPS layer (as the reaction space) rather than in the bulk [32].

Wang et al. [43] studied the function and role of EPS in the bacterial attachment to the cathodic material of waste EV LIBs, a formation of aggregates (cell-EPS-cathode), transfer of electron inside the aggregates, and metals solubilization from the aggregates. It was shown that adherence of bacteria to the spent LIBs particles occurs mediated by EPS layer through a minor electrostatic force and a major hydrophobic force. Furthermore, the EPS adsorb both the ferric ions and more strongly ferrous ions to concentrate the Fe^{2+}/Fe^{3+} cycle inside the aggregates which causes a more powerful reductive attack on metals. The authors proved that except lithium, the effective solubilization of cobalt, nickel, and manganese was essential via direct contact between the spent cathodes and microbial cells.

7.1.2 Non-contact Bioleaching Mechanisms (or Indirect Bioleaching Mechanisms)

In the mechanism of indirect bioleaching, no physical contact is required between solid particles and microbes [64]. The indirect mechanism involves the Fe^{2+}/Fe^{3+} cycle by iron-oxidizing bacteria and bio-oxidation of elemental sulfur by sulfur-oxidizing bacteria and thus formation of inorganic sulfuric acid. The electron donors of ferrous iron and elemental sulfur (S^0) are added externally to the medium as they are not present in LIBs [65]. Two indirect mechanisms cause metals dissolution: (1) acidic and oxidizing attack by H_2SO_4 and (2) intermediate reducing compounds attack [26].

Sulfur-oxidizing bacteria metabolize S^0 to H_2SO_4 via bacterial attack and then the protons generated from this metabolite can attack or react with the metals existing in LIBs [8]. At first, crystals of sulfur are converted to colloidal sulfur and then entered to the cells periplasmic space which is caused maximum bio-sulfuric acid productivity under a long time [26]. Bio-sulfuric acid leaches out the metals by the mechanism of proton attack and redoxlysis according to Eqs. (31) and (32) [7, 24]. Biswal et al. [7] used *Acidithiobacillus thiooxidans* for Li and Co removal from spent LIBs. The authors stated that the spent LIBs contain no sulfidic materials; hence, the metal mobilization could be owing to the indirect mechanism [7]. As an example, the leaching of $LiCoO_2$ by *Acidithiobacillus thiooxidans* can tack place as Eq. (33). It was reported that faster dissolution of Li^+ and Co^{2+} from $LiCoO_2$ is related to higher consumption of protons [4].

$$S^0 + H_2O + \frac{3}{2}O_2 \rightarrow H_2SO_4 \quad (31)$$

$$H_2SO_4 + MO \rightarrow MSO_4 + H_2O \quad (M = \text{Metal}) \quad (32)$$

$$4LiCoO_2 + 3H_2SO_4 \rightarrow Co_3O_4(s) + 2Li_2SO_4(aq) + CoSO_4(aq) + 3H_2O + \frac{1}{2}O_2 \quad (33)$$

During the bioleaching process, the produced protons via bio-oxidation of elemental sulfur were balanced by the consumed protons via the oxidation of the metals [65]. The LIBs homogeneous structure can be destroyed by using H_2SO_4 produced in bioleaching system. Therefore, destruction of crystal structure causes the much easier oxidation and dissolution of metallic phases of waste [66].

Iron-oxidizing bacteria catalyze the oxidation of ferrous iron to ferric ions as shown in Eq. (34) [33]. Based on Eqs. (35–37), reducing reactions of Fe^{3+} ions lead to the formation of protons (H^+ ions) which increase metals recovery [67].

$$4Fe^{2+} + O_2 + 4H^+ \xrightarrow{\text{Microorganism}} 4Fe^{3+} + 2H_2O \quad (34)$$

$$Fe^{3+} + H_2O \rightarrow Fe(OH)^{2+} + H^+ \quad (35)$$

$$Fe^{3+} + 2H_2O \rightarrow Fe(OH)_2^+ + 2H^+ \quad (36)$$

$$Fe^{3+} + 3H_2O \rightarrow Fe(OH)_3 + 3H^+ \quad (37)$$

To solubilize the metals, the bacteria act as follows: Fe^{3+} oxidizing agents attack the samples, react with the metal compound, and convert to Fe^{2+} ions. Then, the bacteria oxidize the Fe^{2+} ions for gaining energy, and when Fe^{2+} oxidize, Fe^{3+} regenerate. Metals solubilization can occur through this cycle between Fe^{2+} and Fe^{3+} [51]. Furthermore, as previously mentioned, the protons generated by iron-oxidizing bacteria attack the oxygen atoms in the metal-oxide component. Then, the protonated atoms of oxygen hydrolyze and the metals detach from the solid matrix [68].

It's worth noting that during the bioleaching process, the formation of hydroxyl sulfate precipitation as jarosite ($MFe_3(SO_4)_2(OH)_6$, where $M = Na^+, Ag^+, K^+, NH_4^+$ or H_3O^+) affects the bioleaching and also reducing metal extraction as shown in Eq. (38) [29].

$$3Fe^{3+} + K^+ + 2SO_4^{2-} + 6H_2O \rightarrow KFe_3(SO_4)_2(OH)_6 + 6H^+ \quad (38)$$

In the case of cathode $LiNi_xCo_yMn_{1-x-y}O_2$, the metal Li is in the form of soluble Li^+ ion and Ni, Co, and Mn are in the form of insoluble high valence states. The conversion of the insoluble high valence Ni^{3+}, Mn^{4+}, Co^{3+} into the soluble low valence Ni^{2+}, Mn^{2+}, Co^{2+} occur through a series of biochemical reactions caused by Fe^{2+}/Fe^{3+} cycle (Eqs. (39)–(41)). The Li releases via direct acid dissolution and other metals mobilization including nickel, cobalt, and manganese is based on a coupling reaction (acid dissolution and reduction) which consumes a huge amount of protons resulting in the increase of pH values and reduction of ORP [2]:

$$Li_2O + 2H^+ \rightarrow 2Li^+ + H_2O \quad (39)$$

$$X^{4+}/X^{3+} + 2e/e \rightarrow X^{2+}(XO) \quad (X = Mn,\ Co,\ Ni) \quad (40)$$

$$XO + 2H^+ \rightarrow X^{2+} + H_2O \quad (41)$$

In addition, the biogenic ferrous ions can reduce Mn^{4+} (insoluble form) into Mn^{2+} (soluble form) in the presence of protons (H^+ ions) through chemical reactions as shown in Eq. (42) [22]:

$$MnO_2 + 2Fe^{2+} + 4H^+ = Mn^{2+} + 2Fe^{3+} + 2H_2O \quad (42)$$

In the case of cathode $LiCoO_2$, the leaching of Co and Li can take place as Eq. (43) by the action of *Acidithiobacillus ferrooxidans* [5].

$$2LiCoO_2 + 6H^+ \rightarrow 2Co^{2+} + 2Li^+ + \frac{1}{2}O_2 + 3H_2O \quad (43)$$

Mishra et al. [8] used a bioleaching method of waste LIBs with cathodic material of $LiCoO_2$ by *Acidithiobacillus ferrooxidans.* The authors used the ferrous ion and elemental sulfur as an energy source for ferric ion and sulfuric acid production in the leaching medium. Results revealed a faster dissolution of Co than Li.

Ijadi et al. [67] reported that based on the mechanism of discharging–recharging reactions of battery, initial NiO(OH) turned to $Ni(OH)_2$. The reaction between $Ni(OH)_2$ and H^+ ions, which are generated by *Acidithiobacillus ferrooxidans*, occurred according to Eq. (44) leading to dissolution of Ni.

$$Ni(OH)_2 + 2H^+ \rightarrow Ni^{2+} + 2H_2O \quad (44)$$

In the other work, Naomi J. Boxall et al. [27] demonstrated the indirect non-contact mechanism of LIBs waste for metals extraction by using biogenic lixiviants of ferric iron and H_2SO_4 in a batch method and sequential batch for mitigating potential issues related to surrounding acid and/or consumption of reagent. A single leaching stage only facilitated low yields (<10%), but multiple sequential leaching stages (4 × 1 h) improved yields of all metals. Biogenic Fe^{3+} leaching with 100 mM sulfuric acid was further enable for the maximum recoveries (81.8% Mn, 74.4% Cu, 60.0% Li, 53.2% Co, and 48.7% Ni).

7.1.3 Determination of Contact and Non-contact Bioleaching Mechanisms

By using the dialysis bag test, which could block the macro-molecules transport with a molecule weight of 8000–15,000, the contact and non-contact mechanisms could be determined. The bacteria could not go through the dialysis bag, but small ions (e.g. Fe^{2+}, Fe^{3+}, and H^+) could freely transport. In this method, the bacterial cells hindered from physical contact with the powder materials and thus the contact mechanism of bioleaching could not occur. By comparing bioleaching recovery of metals between the encapsulated powder and free one, the mechanisms of non-contact or contact bioleaching for metals were determined.

Xin et al. [44] investigated the bioleaching of Li and Co from waste LIBs by mixed culture of iron-oxidizing and sulfur-oxidizing bacteria. They observed that both of the metals had not dissolution differences in trapped and not trapped conditions. The results illustrated that the Co and Li mobilization was based on the non-contact mechanism.

In the other study, Xin et al. [2] investigated the Ni, Mn, Li, and Co bioleaching of three typical cathodes consisting of $LiNi_xCo_yMn_{1-x-y}O_2$, $LiMn_2O_4$, and $LiFePO_4$ from spent EV LIBs. In contrary with the presented results by Xin et al. [44], in this study a dramatically decline in recovery yield for Ni (38.3% to 15%), Co (43.5–25.0%), and Mn (93–42%) and no reduction in Li mobilization were observed when the solid waste was encapsulated in dialysis bag in comparison with the free one. Therefore, a non-contact mechanism for lithium mobilization and a contact mechanism for Co, Ni, and Mn mobilization were required.

7.1.4 Bioleaching Thermodynamics of Metals from the Spent LIBs

The bioleaching mechanism of metals from the cathodic material ($LiCoO_2$) of waste LIBs using FeS_2 and S^0 as mixed energy substrates was shown as the following equation:

$$10O_2 + S + 2FeS_2 + 2LiCoO_2 \rightarrow Li_2SO_4 + 2CoSO_4 + Fe_2(SO_4)_3 \quad (45)$$

While the physical-chemical leaching mechanisms of metals from the cathodic material ($LiCoO_2$) of waste LIBs in the presence of $FeSO_4 + H_2SO_4/H_2O_2$ were shown as the following equations:

$$2FeSO_4 + 2LiCoO_2 + 4H_2SO_4 \rightarrow Fe_2(SO_4)_4 + 2CoSO_4 + Li_2SO_4 + 4H_2 \quad (46)$$

$$3H_2O_2 + 2LiCoO_2 + 3H_2SO_4 \rightarrow 2O_2 \uparrow + 2CoSO_4 + Li_2SO_4 + 6H_2O \quad (47)$$

Table 3 shows the comparison of the thermodynamic parameters of metals bioleaching from waste LIBs using FeS_2 and S^0 as mixed energy substrates with those of the traditional physical–chemical leaching using $FeSO_4 + H_2SO_4/H_2O_2$.

Bioleaching has a huge negative ΔG value. ΔG of bioleaching is 11.4/12.7 times greater than ΔG of the traditional chemical leaching using $H_2O_2/FeSO_4$, respectively. This indicates that although the bioleaching has slow kinetics process, it has much more potential to happen spontaneously than traditional chemical leaching. Bioleaching has huge negative values of both ΔS and ΔH which show that bioleaching is an exothermic chemical reaction. The results show that greater temperatures are not conducive to the reaction. Also, with an increase of reaction temperature, the slight decrease in ΔG values are observed. However, the activity of bioleaching microorganism mostly affected bioleaching performance. The growth and activity of bioleaching mesophilic microorganisms are improved when the temperature rises from 30 to 35 °C leading to enhanced bioleaching yields of lithium and cobalt. While the temperature enhancement from 35 to 40 °C has a negative effect on growth rate and activity of bioleaching mesophilic microbes which results in lower recovery yield of cobalt and lithium [69].

7.1.5 Operational Parameters in Bacterial Leaching of LIBs

Nowadays, biological treatments are less introduced than chemical processes, probably owing to this fact that bioleaching has a lot of parameters which should be adjusted compared to chemical processes [38]. Bioleaching efficiency depends on different parameters such as pulp density, particle size, type of the waste, temperature, pH, ORP, CO_2, and O_2 concentrations in the system [7]. In the following, some of these parameters are described.

pH

The pH adjustment enhances the bacterial growth and activity; thus, the production of bio-sulfuric acid and cycle of Fe^{2+}/Fe^{3+} are accelerated [2]. In the leaching system using Fe^{2+}/Fe^{3+} cycle and pH ranges above 2.0–2.2, the jarosite formation (precipitated at pH $\approx$ 3) occurs which removes Fe^{3+} ions (oxidant) from the leaching solution leading to a drop in bioleaching yields [70].

Table 3 Comparison of the thermodynamic parameter variations between bioleaching processes and chemical leaching using $FeSO_4/H_2O_2$ under different reaction temperatures [69][a]

T (K)	The traditional chemical leaching ($FeSO_4$ as deoxidizer)			The traditional chemical leaching (H_2O_2 as deoxidizer)			The current bioleaching		
	ΔH (KJ/mol)	ΔS (KJ/mol)	ΔG (KJ/mol)	ΔH (KJ/mol)	ΔS (KJ/mol)	ΔG (KJ/mol)	ΔH (KJ/mol)	ΔS (KJ/mol)	ΔG (KJ/mol)
298	−379.04	−381.21	−265.44	−272.13	71.54	−293.45	−4534.96	−3037.01	−3629.93
303			−263.53			−293.81			−3614.74
308			−261.63			−294.52			−3584.37
313			−259.72			−294.52			−3584.37

ΔS and ΔH are independent parameters of temperature
[a]298 K means standard state

It was reported that a correlation exists between the pH value and lithium mobilization in bioleaching system. In the sulfur leaching system which has the lowest pH values, the Li recovery was the highest [44]. Also, in reduced conditions and at high pH values, Co is very soluble but its solubility decreases at low pH values unless the ORP increases concurrently [27].

Pulp Density

Pulp density as an important parameter for determining commercial applications of bioleaching is greatly affected on bioreactor design, operational cost, media consumption and subsequent separation [14, 71]. For example, increasing in waste load from 10 to 20 (g/L) leads to a sharp reduction about 50% in size of the bioreactor and also the quantity of media that results in a great reduction in the cost of bioleaching process [69]. While the low pulp density decreases the economic feasibility and increases the capital cost of the process [27].

The much more toxicity of LIBs powder to the microbial consortium at higher pulp density results in inhibition and endangering the bacterial growth and their activity. In this regard, lower activities of sulfur-oxidizing bacteria cause less generation of biogenic acid and lower activities of iron-oxidizing bacteria lead to less amount of ferric iron [44]; thus, the decline in efficiency of bioleaching occurs for lithium and cobalt recovery as the pulp density increases [69]. Furthermore, pulp density affects alteration behavior of ORP and pH. Generally, the pH and ORP rises and drops, respectively owing to huge consumption of protons for solubilization of lithium and cobalt [69].

It's worth noting that in high pulp density, the O_2 and CO_2 dissolution and their transfer to the bacteria is low leading to enhanced growth of the bacteria and bioleaching performance [72].

Niu et al. [69] investigated the bioleaching of metals from LIBs at different pulp densities. The results showed when pulp density rose from 1 to 4% (w/v), the recovery yields for Co dropped from 52% to 10% and for Li decreased from 80% to 37%. Nevertheless, the high leaching efficiency for lithium and cobalt was 89 and 72%, respectively at a pulp density of 2% (w/v).

Particle Size

One of the challenges in process of chemical engineering especially in bioleaching is mass transfer. Contact surface affects the mass transfer as an important factor. In the bioleaching process, the tension between the surface of particles can be created through the solid pulp that may cause the cell wall lysis. By reduction in the particle size, simultaneously increment of tension and mass transfer occurs due to the increase of the contact surface. Therefore, these two factors in bioleaching of LIBs would balance by sieving the spent battery powder via the right mesh size [43].

Ijadi et al. [67] demonstrated that general formula of metallic hydroxides is $(M(OH)_y)_n$ where M is responsible for metals like nickel and cobalt and n is indicative the metallic hydroxides number in a crystal structure. For metals with a higher value of n, the increase in accessible surface area results from the increase in the size of the powder. Accordingly, they showed that as the particle size decreased, the Ni and Co recovery increased.

Nutrients

Microorganisms need different nutrients in bioleaching process. Marcincakova et al. [4] performed the Li and Co bioleaching from spent LIBs consisting ~ 27.5% $LiCoO_2$. The experiments were performed by *Acidithiobacillus ferrooxidans* and *Acidithiobacillus thiooxidans* in two different media: a rich nutrient medium with all minerals needed for growth of bacteria and a low nutrient medium containing only H_2SO_4 and S^0 as an energy source. In the first medium, the overall Li and Co recovery was 80% and 67%, respectively, but in the second medium, only 35% of Li and 10.5% of Co were recovered. Therefore, the presence of nutrients in the medium of bioleaching influenced Li and Co mobilization from waste LIBs.

It's worth noting that bioleaching of LIBs powder is difficult because they have low content of sulfur and iron which are essential energy sources for acidophilic growth and their activity; hence, supplementation of soluble Fe^{2+} and also S^0 and artificially improving the growth conditions are needed for microbial growth facilitation rate and activity [27].

The concentration of mixed energy substrates (e.g. sulfur and pyrite) affected the performance of waste LIBs bioleaching to some degree. Niu et al. [69] examined the release of cobalt and lithium from spent LIBs using iron-oxidizing and sulfur-oxidizing bacteria. They reported that by increasing the mixed energy substrate concentration from 2 to 4 (g/L), the bioleaching performance increase from 21 to 32% for Co and from 57 to 62% for Li, respectively. The higher concentration of mixed energy substrates (8 (g/L)) did not cause the higher bioleaching recovery for Li and Co, although the concentration 8 (g/L) harvested the highest value of ORP owing to the highest ferric generation using the iron-oxidizing bacteria and the lowest value of pH owing to the highest H_2SO_4 generation by the sulfur-oxidizing bacteria.

Xin et al. [44] studied the bioleaching of metals from waste LIBs using a mixed culture of iron-oxidizing and sulfur-oxidizing bacteria. Bio-acidic mobilization of metals was carried out in three bioleaching systems: S^0, FeS_2, and also $S + FeS_2$ leaching systems. The maximum recovery of lithium was in the S^0 leaching system owing to its lowest pH value of 1.54. The lowest recovery of Li obtained at the highest pH value of 1.69 along with FeS_2 leaching system. It is proposed that the main mechanism for lithium recovery is acid dissolution and independent of energy sources types. For Co mobilization in the S^0 leaching system, the sole mechanism was the acid dissolution, while a combined action of acid dissolution and Fe^{2+} accounted for Co mobilization in the FeS_2 and $S + FeS_2$ systems. The highest recovery of cobalt obtained at higher pH value and varied ORP in the $S + FeS_2$ leaching system and the lowest recovery of cobalt occurred at almost unchanged ORP in the S^0 leaching system.

8 Mixed Culture Bioleaching

In recent years, a number of researchers have performed mixed culture communities for the process of bioleaching [29]. By using the mixed culture of heterotrophs and autotrophs, the synergistic effects occur between microorganisms. Mixed culture bioleaching can be more effective in metal solubilization compared to pure cultures of single-strain [30]. In the future, more research will be accomplished by mixed cultures communities owing to its potential to fabricate process of bioleaching more practical [34].

Mixed culture of *Acidithiobacillus thiooxidans* and *Acidithiobacillus ferrooxidans* demonstrates a more efficient metal mobilization than separated culture. In the mixed culture, the desired acidic conditions for growth of *Acidithiobacillus ferrooxidans* is created by *Acidithiobacillus thiooxidans* through producing H_2SO_4 [29]. It was reported that at the initial leaching stage, the pH values of culture medium decreased more sharply with the mixed culture compared to that for the pure culture. Also, by mixed culture, the solution redox potentials reached faster to ultimate potentials (range of 550–600 mV) in comparison with pure culture [73].

It was reported that a mixed culture exhibits a higher recovery of manganese (insoluble Mn^{4+} to soluble Mn^{2+}) by Fe^{2+} in comparison with a pure culture of *Acidithiobacillus thiooxidans*. In mixed culture system, the reduction is reflected by the great fluctuation in ORP during the bioleaching process. In addition, the lower pH value in mixed culture causes stronger production of ferric ions than single-strain iron-oxidizing-bacteria. Indeed, more ferric ions lead to a higher generation of ferrous ions which highly accelerate the reduction of Mn^{4+} ions and also the recovery of Mn^{2+}. Hence, the use of mixed culture results in the maximum manganese recovery [22, 44].

Heydarian et al. [29] investigated two-step bioleaching of laptops LIBs by using a mixture of *Acidithiobacillus thiooxidans* and *Acidithiobacillus ferrooxidans*. The role of bacteria in a mixed culture system is maintaining the ORP of a system in a high value via balancing the Fe^{2+}/Fe^{3+} cycle, bio-sulfuric production for providing H^+ ions, and also preserving a high concentration of ferric ions. The bio-acid of *Acidithiobacillus thiooxidans* also helped to maintain the acidity of the medium. The ratio of both strains in mixed culture is an important factor for the pH balancing in the mechanisms of solubility of each metal. The inoculum ratio of 3:2 for *Acidithiobacillus ferrooxidans* and *Acidithiobacillus thiooxidans* was the best obtained ratio. The maximum recovery of metals reached to 99.2% of lithium, 50.4% of cobalt and 89.4% of nickel under optimum conditions of initial pH of 1.5, 5.0 (g/L) elemental sulfur, and 36.7 (g/L) iron sulfate.

It can be possible to perform bioleaching by a combination of bio-metabolites. It was reported that bioleaching by combination of the biogenic acid and ferric ion enhanced the leaching performance compared to the individual bio-reagents in terms of pH and ORP. The pH value of combined bio-reagents was more acidic than the single bio-reagents, which means more oxidizing conditions leading to more metal solubilization [27].

9 Application of Catalyst in Bioleaching

Although bioleaching has many advantages, its slow kinetics for metals dissolution is a critical parameter that hinders its application to be economical for large-scales [33]. To improve the rate of metal dissolution and bioleaching yield, the use of suitable catalysts can be considered. A "catalyst" is a matter which decreases the activation energy and thus increases the reaction rate. A variety of catalytic metallic ions such as Hg^{2+}, Co^{2+}, Ag^{+}, Cu^{2+}, Bi^{3+}, etc. and non-metallic catalysts such as light illumination, polyethylene glycol, activated carbon, etc. have been considered for improving the efficiency of bioleaching process. Two drawbacks of non-metallic catalysts are poor bioleaching yields and need of large catalyst amounts for achieving significant recovery yields. Among different catalysts, the application of metal ions has attracted the most attention. It is expected that in a near future, metal ions affect significantly on the development of large-scale bioleaching technology due to their excellent catalytic properties [33]. Catalytic metallic ions can speed up electron transfer and thus improve both the performance of metal extraction and substrate oxidation [14]. There is a common view on the catalytic effect of metal ions on bioleaching processes: metal intermediate formation on the particles surface which dissolves easily leading to improvement of bioleaching performance [5].

Among different metal ions, silver ions (Ag^{+}) gain the most attention. In the bioleaching process, several compounds of silver such as silver nitrate ($AgNO_3$), silver chloride (AgCl), and silver sulfate (Ag_2SO_4) have been applied as a source of Ag^{+}. It was reported that insoluble silver salts (e.g. Ag_2S) can be applied instead of soluble ones [33]. In the case of LIBs bioleaching by *Acidithiobacillus ferrooxidans*, the mechanism of catalytic action of Ag^{+} for the $LiCoO_2$ bioleaching can be illustrated as [5]:

$$Ag^{+} + LiCoO_2 \rightarrow AgCoO_2 + Li^{+} \tag{48}$$

$$AgCoO_2 + 3Fe^{3+} \rightarrow 3Fe^{2+} + Ag^{+} + O_2 + Co^{2+} \tag{49}$$

$$4Fe^{2+} + O_2 + 4H^{+} \xrightarrow{A.ferrooxidans} 4Fe^{3+} + 2H_2O \tag{50}$$

The key catalytic role of Ag^{+}, as shown in Eq. (48), is the facilitation in $AgCoO_2$ intermediate formation. As shown in Eq. (49), when the Fe^{3+} ions, which produce by the action of bacteria *Acidithiobacillus ferrooxidans*, oxidize $AgCoO_2$ intermediate, the consumed Ag^+ is released which can be used for further catalytic action. The produced Fe^{2+} ions are oxidized back to Fe^{3+} ions because of bacterial activity (Eq. (50)) [5].

For the first time, Zeng et al. [5] reported the application of Ag-catalyzed bacteria (*Acidithiobacillus ferrooxidans*) for cobalt dissolution from spent LIBs. The authors reported that addition of only 0.02 (g/L) Ag^+ ions to the culture medium led to 98.4% dissolution of Co within 7 days, whilst in the absence of Ag^+, the dissolution of Co was only 43.1%.

Application of other catalytic ions for developing an affordable bioleaching process has been tried. Use of Cu^{2+} ions has been considered owing to its easy availability and relatively cheaper cost [33].

Zeng et al. [41] studied the effect of Cu^{2+} ions on cobalt dissolution from spent LIBs using *Acidithiobacillus ferrooxidans*. Dissolution of 99.9% of Co in the presence of 0.75 (g/L) Cu^{2+} occurred after 6 days, whilst only 43.1% of Co was dissolved after 10 days in the absence of Cu^{2+}. The authors reported that the catalytic action of copper ions was owing to the $CuCo_2O_4$ intermediate formation on the surface of $LiCoO_2$ caused by cationic exchange reactions. The Fe^{3+} ions could easily dissolve the intermediate product ($CuCo_2O_4$), resulting in the higher Co recovery yield as shown in the following equations:

$$Cu^{2+} + 2LiCoO_2 \rightarrow CuCo_2O_4 + 2Li^+ \tag{51}$$

$$CuCo_2O_4 + 6Fe^{3+} \rightarrow 6Fe^{2+} + Cu^{2+} + 2O_2 + 2Co^{2+} \tag{52}$$

From the aspect of an equal mass of Ag and Cu, the Ag-catalyzed bacteria can achieve higher bioleaching efficiency compared with the Cu-catalyzed bacteria [5].

However, using metal ion catalysts have several key challenges and in spite of its high effectiveness, the development of this process has not occurred for larger scale applications. One of the challenges is the high-cost catalytic metals such as Hg, Bi, Ag, and Ru that needs to find a more effective and cheaper metal catalyst [41]. Another challenge is related to the potential inhibitory/toxic effects of the catalytic metals to microorganisms. For example, it was reported that low amount of 0.1 (mg/L) of Ag^+ can inhibit the iron oxidation rate of *Acidithiobacillus ferrooxidans*. In the bacterial cell, the Ag^+ ions can replace Fe^{2+} ions in the active site of the enzyme with oxidizing ability. With regard to these challenges, the identification and employing the resistant microorganism with the ability to tolerate a higher concentration of metallic ions is very important [33].

10 Application of Ultrasound in Bioleaching (Sonobioleaching)

For improving the slow kinetics of bioleaching and overcoming this drawback, different strategies have been used. One of these approaches is using ultrasound in bioleaching processes which named sonobioleaching. Ultrasonic cavitation in the form of micro-jet and shock waves formation increases the rate and dissolution of metals in the leaching reactions using metabolites produced by microbes [74]. In the ultrasound process, a higher enhancement in the leaching rate occurs at low concentrations of acid; this matter is beneficial for bioleaching method due to the production of low metabolite concentration in bio-processes [75]. The ultrasonic process causes the easier penetration of bio-lixiviant to the solid particles [74].

The ultrasound can affect the microbial metabolic activity which relies on several parameters such as the type of microbes, their size and shape, and environmental parameters including temperature, pressure, ionic strength, and acid/base concentration [75]. It has been reported that sonobioleaching increases the microbial/enzymatic activities due to stirring improvement at macroscopic and microscopic levels and changing the cell membranes permeability [34].

Vargas et al. [76] achieved 28% enhancement in metabolic activity through sonication of more than 4 min owing to the *Aspergillus niger* disruption and invertase enzymes release.

Improving the bioleaching efficiency through sonobioleaching depends on the type of bioleaching approach including one-step, two-step, and spent-medium bioleaching [75]. It has been reported that low-frequency ultrasound leads the growth enhancement while high-frequency ultrasound causes cell walls disruption [76]. An optimum frequency of 40 kHz at the intensity of 1.5 W was reported to increase the growth of *Aspergillus niger*; below and above this threshold, the ultrasound effectiveness was lower [74].

However, the ultrasonic application in metals recovery from spent LIBs has not been investigated in detail yet.

11 A Brief Review of Previous Studies on LIBs Bioleaching

Notwithstanding the growing number of studies on LIBs bioleaching, its application in the large-scales is still limited. Table 4 demonstrates a summary of previous studies on bioleaching of LIBs including the applied technique, microorganisms, condition, and recovery of metals.

Table 4 Studies in the bioleaching of spent LIBs

Method	Agent	Recovery (%)	Condition	References
One-step bioleaching	Consortia of isolated bacteria	Li: 62.8	2 (mg/mL) pulp density, 30 °C, 120 rpm, 15 days	[11]
One-step bioleaching	*Acidithiobacillus ferrooxidans* *Acidithiobacillus thiooxidans*	Co: 67 Li: 80	1% (w/v) pulp density	[4]
One-step bioleaching	*Acidithiobacillus ferrooxidans*	Co: 65 Li: 10	5 (g/L) pulp density, 1% elemental sulfur, 3 (g/L) Fe(II), 30 °C, 200 rpm	[8]
Two-step bioleaching	*Acidithiobacillus thiooxidans* *Leptospirillum ferriphilum*	Li: 98	1% (w/v) pulp density, $LiFePO_4$ cathode, 30 °C, 120 rpm	[2]
		Li: 95 Mn: 96	1% (w/v) pulp density, $LiMn_2O_4$ cathode, 30 °C, 120 rpm	
		Li: Mn: > 90 Co: Ni: 40	1% (w/v) pulp density, $LiNi_xCo_yMn_{1\_x\_y}O_2$ cathode, 30 °C, 120 rpm	
Two-step bioleaching	*Acidithiobacillus ferrooxidans* *Acidithiobacillus thiooxidans*	Li: 99.2 Ni: 89.4 Co: 50.4	40 (g/L) pulp density, 5.0 (g/L) sulfur, 36.7 (g/L) iron sulfate, 32 °C, 160 rpm, 16 days	[29]
Two-step bioleaching	*Alicyclobacillus* sp. *Sulfobacillus* sp.	Li: 89 Co: 72	20 (g/L) pulp density, 30 °C	[69]
Two-step bioleaching	*Acidithiobacillus ferrooxidans*	Co: 99.9	1% (w/v) pulp density, 0.75 (g/L) copper ions as catalyst, 35 °C, 160 rpm, 6 days	[41]
Two-step bioleaching	*Acidithiobacillus ferrooxidans*	Co: 98.4	1% (w/v) pulp density, 0.02 (g/L) silver ions as catalyst, 35 °C, 160 rpm, 7 days	[5]

(continued)

Table 4 (continued)

Method	Agent	Recovery (%)	Condition	References
Spent-medium bioleaching	*Acidithiobacillus ferrooxidans* *Acidithiobacillus thiooxidans*	Li: 60.0 Ni: 48.7 Mn: 81.8 Cu: 74.4 Co: 53.2	10% (w/v) pulp density, 22 °C, Biogenic ferric with 100 (mM) H_2SO_4, 1 h × 4 cycles	[27]
Spent-medium bioleaching	*Aspergillus niger*	Li: 95 Co: 45 Cu: 100 Mn: 70 Al: 65 Ni: 38	1% (w/v) pulp density, 30 °C, 130 rpm, 16 days	[3]
Spent-medium bioleaching	*Aspergillus niger*	Li: 100 Cu: 100 Mn: 77 Al: 75	2% (w/v) pulp density, 30 °C, 130 rpm, 8 days	[9]
		Co: 64 Ni: 54	1% (w/v) pulp density, 30 °C, 130 rpm, 8 days	
Spent-medium bioleaching	*Aspergillus niger* MM1 and SG1 strains	Co: 82 Li: 100	0.25% (w/v) pulp density, 30 °C, 200 rpm, 40 days	[7]
	Acidithiobacillus thiooxidans	Co: 22 Li: 66	0.25% (w/v) pulp density, 30 °C, 200 rpm, 40 days	

12 Future Prospects in Bio-Hydrometallurgical Methods

The existing status of bio-hydrometallurgical approach in the worldwide scenario is encouraging. Today, a bio-hydrometallurgical approach is no longer a promising future technology but an effective and actual economic and environmental alternative for metals recovery from secondary wastes. There is an ever growing acceptance of bio-hydrometallurgical approach as a commercially feasible technology that has already given revolutionary solutions to the historical problems related to traditional methods of metals recovery.

However, at the present time, large-scale application of bioleaching from LIBs does not exist and its technology remains at the laboratory scale. It is anticipated that in near future, new industrial microbial metal leaching plants will be installed which require more specific researches to develop their commercial prospects.

Major challenges exist in bioleaching processes to be applied commercially including the slow process kinetics, waste load (ratio of solid to liquid), application of new approaches for strain improvement to increase the metabolites production, the design of simple bioreactors for the microbial cultivation and find a cheap nutrient source for microbes. In this regard, the advanced knowledge about the interactions between wastes (e.g. spent LIBs) and microbes is needed. Altering of the $LiCoO_2$ structure by pretreatments including ultrasound, microwave, and heat and also use the hybrid leaching including a combination of both acidophilic autotrophic and heterotrophic microorganisms are alternatives to improve spent LIBs commercial bioleaching performance in the future. Furthermore, for enhancing metals recovery, new microbial strains with higher ability to tolerate toxic metals should be identified. However, the challenges of biological processes demand close cooperation between engineers, microbiologists, and chemists.

Acknowledgements This study was financially supported by Tarbiat Modares University under grant number IG-39701.

References

1. J.H. Duffus, " Heavy metals" a meaningless term? (IUPAC technical report). Pure Appl. Chem. **74**, 793–807 (2002)
2. Y. Xin, X. Guo, S. Chen, J. Wang, F. Wu, B. Xin, Bioleaching of valuable metals Li Co, Ni and Mn from spent electric vehicle Li-ion batteries for the purpose of recovery. J. Clean. Prod. **116**, 249–258 (2015)
3. N. Bahaloo-Horeh, S.M. Mousavi, S.A. Shojaosadati, Bioleaching of valuable metals from spent lithium-ion mobile phone batteries using *Aspergillus niger*. J. Power Sour. **320**, 257–266 (2016)
4. R. Marcincakova, J. Kadukova, A. Mrazikova, O. Velgosova, A. Luptakova, S. Ubaldini, Metal bioleaching from spent lithium-ion batteries using acidophilic bacterial strains. Inżynieria Miner **17**, 117–120 (2016)
5. G. Zeng, S. Luo, X. Deng, L. Li, C. Au, Influence of silver ions on bioleaching of cobalt from spent lithium batteries. Miner. Eng. **49**, 40–44 (2013)
6. M. Sun, Y. Wang, J. Hong, J. Dai, R. Wang, Z. Niu, B. Xin, Life cycle assessment of a bio-hydrometallurgical treatment of spent Zn-Mn batteries. J. Clean. Prod. **129**, 350–358 (2016)
7. B.K. Biswal, U.U. Jadhav, M. Madhaiyan, L. Ji, E.-H. Yang, B. Cao, Biological leaching and chemical precipitation methods for recovery of Co and Li from spent lithium-ion batteries. ACS Sustain. Chem. Eng. **6**, 12343–12352 (2018)
8. D. Mishra, D.-J. Kim, D.E. Ralph, J.-G. Ahn, Y.-H. Rhee, Bioleaching of metals from spent lithium ion secondary batteries using *Acidithiobacillus ferrooxidans*. Waste Manage. **28**, 333–338 (2008)
9. N. Bahaloo-Horeh, S.M. Mousavi, Enhanced recovery of valuable metals from spent lithium-ion batteries through optimization of organic acids produced by *Aspergillus niger*. Waste Manage. **60**, 666–679 (2017)
10. A.D. Zand, M.A. Abduli, Current situation of used household batteries in Iran and appropriate management policies. Waste Manage. **28**, 2085–2090 (2008)
11. M. Hartono, M.A. Astrayudha, H. Petrus, W. Budhijanto, H. Sulistyo, Lithium recovery of spent lithium-ion battery using bioleaching from local sources microorganism. Rasayan J Chem **10**, 897–903 (2017)
12. W. Zhang, C. Xu, W. He, G. Li, J. Huang, A review on management of spent lithium-ion batteries and strategy for resource recycling of all components from them. Waste Manage. Res. **36**, 99–112 (2018)
13. S. Ubaldini, J. Kadukova, A. Mrazikova, P. Fornari, A. Luptakova, R. Marcincakova, P. Pizzichemi, Application of innovative processes for the valorisation of spent alkaline batteries. Chem. Eng. Trans. **39**, 1609–1614 (2014)

14. Z. Niu, Q. Huang, J. Wang, Y. Yang, B. Xin, S. Chen, Metallic ions catalysis for improving bioleaching yield of Zn and Mn from spent Zn-Mn batteries at high pulp density of 10%. J. Hazard. Mater. **298**, 170–177 (2015)
15. K.M. Winslow, S.J. Laux, T.G. Townsend, A review on the growing concern and potential management strategies of waste lithium-ion batteries. Resour. Conserv. Recycl. **129**, 263–277 (2018)
16. J. Lee, B.D. Pandey, Bio-processing of solid wastes and secondary resources for metal extraction–a review. Waste Manage. **32**, 3–18 (2012)
17. M.-J. Kim, J.-Y. Seo, Y.-S. Choi, G.-H. Kim, Bioleaching of spent Zn–Mn or Ni-Cd batteries by *Aspergillus* species. Waste Manage. **51**, 168–173 (2016)
18. L.-P. He, S.-Y. Sun, X.-F. Song, J.-G. Yu, Leaching process for recovering valuable metals from the $LiNi_{1/3}Co_{1/3}Mn_{1/3}O_2$ cathode of lithium-ion batteries. Waste Manage. **64**, 171–181 (2017)
19. S. Bindschedler, T.Q.T.V. Bouquet, D. Job, E. Joseph, P. Junier, Fungal biorecovery of gold from e-waste. Adv. Appl. Microbiol. Elsevier, 53–81 (2017)
20. N. Bahaloo-Horeh, S.M. Mousavi, M. Baniasadi, Use of adapted metal tolerant *Aspergillus niger* to enhance bioleaching efficiency of valuable metals from spent lithium-ion mobile phone batteries. J. Clean. Prod. **197**, 1546–1557 (2018)
21. J. Dewulf, G. Van der Vorst, K. Denturck, H. Van Langenhove, W. Ghyoot, J. Tytgat, K. Vandeputte, Recycling rechargeable lithium-ion batteries: critical analysis of natural resource savings. Resour. Conserv. Recycl. **54**, 229–234 (2010)
22. B. Xin, W. Jiang, H. Aslam, K. Zhang, C. Liu, R. Wang, Y. Wang, Bioleaching of zinc and manganese from spent Zn–Mn batteries and mechanism exploration. Bioresour. Technol. **106**, 147–153 (2012)
23. A.M. Diederen, *Metal minerals scarcity: a call for managed austerity and the elements of hope* (Saf, TNO Defence, Secur, 2009)
24. A. Akcil, F. Vegliò, F. Ferella, M.D. Okudan, A. Tuncuk, A review of metal recovery from spent petroleum catalysts and ash. Waste Manage. **45**, 420–433 (2015)
25. O. Velgosová, J. Kaduková, R. Marcinčáková, A. Mražíková, L. Fröhlich, The role of main leaching agents responsible for Ni bioleaching from spent Ni-Cd batteries. Sep. Sci. Technol. **49**, 438–444 (2014)
26. C. Cerruti, G. Curutchet, E. Donati, Bio-dissolution of spent nickel-cadmium batteries using *Thiobacillus ferrooxidans*. J. Biotechnol. **62**, 209–219 (1998)
27. N.J. Boxall, K.Y. Cheng, W. Bruckard, A.H. Kaksonen, Application of indirect non-contact bioleaching for extracting metals from waste lithium-ion batteries. J. Hazard. Mater. **360**, 504–511 (2018)
28. O. Velgosová, J. Kaduková, R. Marcinčáková, Study of Ni and Cd bioleaching from spent Ni-Cd batteries. Nov. Biotechnol. Chim. **11**, 117–123 (2012)
29. A. Heydarian, S.M. Mousavi, F. Vakilchap, M. Baniasadi, Application of a mixed culture of adapted acidophilic bacteria in two-step bioleaching of spent lithium-ion laptop batteries. J. Power Sour. **378**, 19–30 (2018)
30. A. Işıldar, Biotechnologies for metal recovery from electronic waste and printed circuit boards. Waste Electr. Electron. Equip. Recycl. Elsevier, 241–269 (2018)
31. F. Beolchini, V. Fonti, A. Dell'Anno, L. Rocchetti, F. Vegliò, Assessment of biotechnological strategies for the valorization of metal bearing wastes. Waste Manage. **32**, 949–956 (2012)
32. D. Mishra, Y.H. Rhee, Microbial leaching of metals from solid industrial wastes. J. Microbiol. **52**, 1–7 (2014)
33. A. Pathak, L. Morrison, M.G. Healy, Catalytic potential of selected metal ions for bioleaching, and potential techno-economic and environmental issues: a critical review. Bioresour. Technol. **229**, 211–221 (2017)
34. T. Gu, S.O. Rastegar, S.M. Mousavi, M. Li, M. Zhou, Advances in bioleaching for recovery of metals and bioremediation of fuel ash and sewage sludge. Bioresour. Technol. **261**, 428–440 (2018)
35. H. Brandl, Microbial leaching of metals. Biotechnol. Set. 191–224 (2008) (Second Ed)

36. M.E. Hoque, O.J. Philip, Biotechnological recovery of heavy metals from secondary sources—An overview. Mater. Sci. Eng. C **31**, 57–66 (2011)
37. D. Santhiya, Y.-P. Ting, Use of adapted *Aspergillus niger* in the bioleaching of spent refinery processing catalyst. J. Biotechnol. **121**, 62–74 (2006)
38. A. Potysz, E.D. van Hullebusch, J. Kierczak, Perspectives regarding the use of metallurgical slags as secondary metal resources–a review of bioleaching approaches. J. Environ. Manage. **219**, 138–152 (2018)
39. I. Asghari, S.M. Mousavi, F. Amiri, S. Tavassoli, Bioleaching of spent refinery catalysts: a review. J. Ind. Eng. Chem. **19**, 1069–1081 (2013)
40. K. Pollmann, S. Kutschke, S. Matys, J. Raff, G. Hlawacek, F.L. Lederer, Bio-recycling of metals: recycling of technical products using biological applications. Biotechnol. Adv. **36**, 1048–1062 (2018)
41. G. Zeng, X. Deng, S. Luo, X. Luo, J. Zou, A copper-catalyzed bioleaching process for enhancement of cobalt dissolution from spent lithium-ion batteries. J. Hazard. Mater. **199–200**, 164–169 (2012)
42. A. Işıldar, E.D. van Hullebusch, M. Lenz, G. Du Laing, A. Marra, A. Cesaro, S. Panda, A. Akcil, M.A. Kucuker, K. Kuchta, Biotechnological strategies for the recovery of valuable and critical raw materials from waste electrical and electronic equipment (WEEE)–A review. J. Hazard. Mater. **362**, 467–481 (2019)
43. J. Wang, B. Tian, Y. Bao, C. Qian, Y. Yang, T. Niu, B. Xin, Functional exploration of extracellular polymeric substances (EPS) in the bioleaching of obsolete electric vehicle $LiNi_xCo_yMn_{1-x-y}O_2$ Li-ion batteries. J. Hazard. Mater. **354**, 250–257 (2018)
44. B. Xin, D. Zhang, X. Zhang, Y. Xia, F. Wu, S. Chen, L. Li, Bioleaching mechanism of Co and Li from spent lithium-ion battery by the mixed culture of acidophilic sulfur-oxidizing and iron-oxidizing bacteria. Bioresour. Technol. **100**, 6163–6169 (2009)
45. A.H. Kaksonen, C. Morris, J. Wylie, J. Li, K. Usher, F. Hilario, C.A. du Plessis, Continuous flow 70 °C archaeal bioreactor for iron oxidation and jarosite precipitation. Hydrometallurgy **168**, 40–48 (2017)
46. F. Anjum, M. Shahid, A. Akcil, Biohydrometallurgy techniques of low grade ores: a review on black shale. Hydrometallurgy **117**, 1–12 (2012)
47. J.D. Garcia-Garcia, R. Sanchez-Thomas, R. Moreno-Sanchez, Bio-recovery of non-essential heavy metals by intra-and extracellular mechanisms in free-living microorganisms. Biotechnol. Adv. **34**, 859–873 (2016)
48. C.J. Liang, J.Y. Li, C.J. Ma, Review on cyanogenic bacteria for gold recovery from e-waste. Adv. Mater. Res. **878**, 355–367 (2014)
49. M. Motaghed, S.M. Mousavi, S.O. Rastegar, S.A. Shojaosadati, Platinum and rhenium extraction from a spent refinery catalyst using *Bacillus megaterium* as a cyanogenic bacterium: statistical modeling and process optimization. Bioresour. Technol. **171**, 401–409 (2014)
50. Y. Lu, Z. Xu, Precious metals recovery from waste printed circuit boards: a review for current status and perspective. Resour. Conserv. Recycl. **113**, 28–39 (2016)
51. M. Arshadi, S.M. Mousavi, Simultaneous recovery of Ni and Cu from computer-printed circuit boards using bioleaching: statistical evaluation and optimization. Bioresour. Technol. **174**, 233–242 (2014)
52. A. Işıldar, J. van de Vossenberg, E.R. Rene, E.D. van Hullebusch, P.N.L. Lens, Two-step bioleaching of copper and gold from discarded printed circuit boards (PCB). Waste Manage. **57**, 149–157 (2016)
53. S. Ilyas, J. Lee, Biometallurgical recovery of metals from waste electrical and electronic equipment: a review. ChemBioEng Rev **1**, 148–169 (2014)
54. H. Brandl, S. Lehmann, M.A. Faramarzi, D. Martinelli, Biomobilization of silver, gold, and platinum from solid waste materials by HCN-forming microorganisms. Hydrometallurgy **94**, 14–17 (2008)
55. K. Pollmann, S. Kutschke, S. Matys, S. Kostudis, S. Hopfe, J. Raff, Novel biotechnological approaches for the recovery of metals from primary and secondary resources. Minerals **6**, 54 (2016)

56. B. Max, J.M. Salgado, N. Rodríguez, S. Cortés, A. Converti, J.M. Domínguez, Biotechnological production of citric acid. Braz. J. Microbiol. **41**, 862–875 (2010)
57. J.-O. Kim, Y.-W. Lee, J. Chung, The role of organic acids in the mobilization of heavy metals from soil. KSCE J. Civ. Eng. **17**, 1596–1602 (2013)
58. Y. Qu, B. Lian, B. Mo, C. Liu, Bioleaching of heavy metals from red mud using *Aspergillus niger*. Hydrometallurgy **136**, 71–77 (2013)
59. L. Li, J. Ge, R. Chen, F. Wu, S. Chen, X. Zhang, Environmental friendly leaching reagent for cobalt and lithium recovery from spent lithium-ion batteries. Waste Manage. **30**, 2615–2621 (2010)
60. L. Sun, K. Qiu, Organic oxalate as leachant and precipitant for the recovery of valuable metals from spent lithium-ion batteries. Waste Manage. **32**, 1575–1582 (2012)
61. S. Ghosh, S. Mohanty, A. Akcil, L.B. Sukla, A.P. Das, A greener approach for resource recycling: manganese bioleaching. Chemosphere **154**, 628–639 (2016)
62. R.-L. Yu, O.U. Yang, J.-X. Tan, F.-D. Wu, S.U.N. Jing, M. Lei, D.-L. Zhong, Effect of EPS on adhesion of *Acidithiobacillus ferrooxidans* on chalcopyrite and pyrite mineral surfaces. Trans. Nonferrous Met. Soc. China **21**, 407–412 (2011)
63. W. Zeng, G. Qiu, H. Zhou, X. Liu, M. Chen, W. Chao, C. Zhang, J. Peng, Characterization of extracellular polymeric substances extracted during the bioleaching of chalcopyrite concentrate. Hydrometallurgy **100**, 177–180 (2010)
64. K. Bosecker, Bioleaching: metal solubilization by microorganisms. FEMS Microbiol. Rev. **20**, 591–604 (1997)
65. A. Marra, A. Cesaro, E.R. Rene, V. Belgiorno, P.N.L. Lens, Bioleaching of metals from WEEE shredding dust. J. Environ. Manage. **210**, 180–190 (2018)
66. L. Zhao, D. Yang, N.-W. Zhu, Bioleaching of spent Ni-Cd batteries by continuous flow system: effect of hydraulic retention time and process load. J. Hazard. Mater. **160**, 648–654 (2008)
67. M. Ijadi Bajestani, S.M. Mousavi, S.A. Shojaosadati, Bioleaching of heavy metals from spent household batteries using *Acidithiobacillus ferrooxidans*: Statistical evaluation and optimization. Sep. Purif. Technol. **132**, 309–316 (2014)
68. F. Gerayeli, F. Ghojavand, S.M. Mousavi, S. Yaghmaei, F. Amiri, Screening and optimization of effective parameters in biological extraction of heavy metals from refinery spent catalysts using a thermophilic bacterium. Sep. Purif. Technol. **118**, 151–161 (2013)
69. Z. Niu, Y. Zou, B. Xin, S. Chen, C. Liu, Y. Li, Process controls for improving bioleaching performance of both Li and Co from spent lithium-ion batteries at high pulp density and its thermodynamics and kinetics exploration. Chemosphere **109**, 92–98 (2014)
70. S.O. Rastegar, S.M. Mousavi, S.A. Shojaosadati, R.S. Mamoory, Bioleaching of V, Ni, and Cu from residual produced in oil fired furnaces using *Acidithiobacillus ferrooxidans*. Hydrometallurgy **157**, 50–59 (2015)
71. B. Xin, W. Jiang, X. Li, K. Zhang, C. Liu, R. Wang, Y. Wang, Analysis of reasons for decline of bioleaching efficiency of spent Zn–Mn batteries at high pulp densities and exploration measure for improving performance. Bioresour. Technol. **112**, 186–192 (2012)
72. S.O. Rastegar, S.M. Mousavi, S.A. Shojaosadati, T. Gu, Bioleaching of fuel-oil ash using *Acidithiobacillus thiooxidans* in shake flasks and a slurry bubble column bioreactor. RSC Adv. **6**, 21756–21764 (2016)
73. H. Liu, G. Gu, Y. Xu, Surface properties of pyrite in the course of bioleaching by pure culture of *Acidithiobacillus ferrooxidans* and a mixed culture of *Acidithiobacillus ferrooxidans* and *Acidithiobacillus thiooxidans*. Hydrometallurgy **108**, 143–148 (2011)
74. F. Anjum, M. Shahid, S. Bukhari, J.H. Potgieter, Combined ultrasonic and bioleaching treatment of hospital waste incinerator bottom ash with simultaneous extraction of selected metals. Environ. Technol. **35**, 262–270 (2014)
75. S. Vyas, Y.-P. Ting, A review of the application of ultrasound in bioleaching and insights from sonication in (bio) chemical processes. Resources **7**, 3 (2018)
76. L.H.M. Vargas, A.C.S. Pião, R.N. Domingos, E.C. Carmona, Ultrasound effects on invertase from *Aspergillus niger*. World J. Microbiol. Biotechnol. **20**, 137–142 (2004)

Impacts of Recycling of Spent Lithium-Ion Batteries on Environmental Burdens

Guangzhe Li and Liang An

Abstract A well-designed recycling process of spent lithium-ion batteries (LIBs) should not only own high recovery efficiency but also be environmentally benign. Most of the recycling of spent LIBs reported exhibited good products recovery, however, the environmental burdens caused by the recycling techniques were far less concerned. In this way, the environmental burdens of the recycling process are specifically addressed in recent years. In this chapter, we gave a systematic overview of this issue, including potential risks of spent LIBs, the assessment methodology and evaluation results of environmental burdens of spent LIBs, the discussion on the environmental impact of recycling process and the approaches for designing an environmentally benign recycling process.

1 Introduction

Since lithium-ion batteries (LIBs) present superiorities in terms of high energy density, they serve as the primary power supply for portable electronic devices and electric vehicles, leading an explosion of LIB industry in the past decade. Large amounts of LIBs were produced, put into market, and finally disposed after usage, which brings great concerns on the environmental impacts of spent LIBs with such a large scale [1–5]. In order to reduce the environmental burden and recycle the high value products from spent LIBs, the recovery techniques of different commercialized spent LIBs have been widely studied [6–15]. The standards for recycling techniques should own high recovery efficiency as well as environmental benignity. The recycling techniques reported were well developed by optimizing conditions in the leaching and precipitation process, which aims to achieve high recovery efficiency. Nevertheless, the impacts of recycling process itself on environmental burden are far less concerned, including the production and consumption of leaching agents, the input energy in recycling and the treatment of byproducts. Therefore, systematic

G. Li · L. An (✉)
Department of Mechanical Engineering, The Hong Kong Polytechnic University, Hung Hom, Kowloon, Hong Kong SAR, China
e-mail: liang.an@polyu.edu.hk

L. An (ed.), *Recycling of Spent Lithium-Ion Batteries*,
https://doi.org/10.1007/978-3-030-31834-5_8

assessment of the environmental impact caused by recycling process is required. Life cycle impact assessment (LCIA) method are frequently used to assess the environmental impacts of LIBs and factors like energy input, environmental toxicity and resources depletion potential can be specifically analyzed by LCIA method [16–26]. In this chapter, we will first discuss the potential risk of the waste of spent LIBs from the view point of the components in spent LIBs, including anode material, cathode material, electrolyte, separator and binder. Then, we will introduce the impact assessment methodology and case studies of spent LIBs evaluation. Finally, we will discuss the impact of recycling process on environment and the possible approaches to reduce the environmental burden.

2 The Potential Risk on Environment of Spent LIBs

The commercial LIBs consist of the following components: cathode material, anode material, electrolyte, separator and binder. Among them, there are chemicals like organic electrolyte, heavy metals and polymer which may directly cause environmental pollution including water pollution, soil pollution and white pollution if disposing casually [2]. Besides, the side reactions of these chemicals may also bring huge risks in disposing and recycling process, which will be summarized as follows.

2.1 Anode Material

As for the commercialized primary LIBs, lithium metal is used as anode material in LIBs to achieve a high energy density. However, if the battery is over-discharging, the lithium dendrit can form and even be molten owing to the releasing heat in discharge. Therefore, improper handling of over-discharging batteries may also pose significant environmental risks. In addition, graphite is commonly used as anode material in LIBs and it is combustible or even explosive, generating huge amounts of carbon dioxide and thus aggravating the greenhouse effect. Not fully discharged LIBs will contain lithium-carbon alloy, which can also react with water to form strong alkaline solution.

2.2 Cathode Material

The type of cathode material and their potential risks are listed in Table 1. Toxic gases may generate if cathode material is treated improperly. The high input energy for material production also leads to environmental burdens.

Table 1 The potential risks on environment of spent cathode materials

Cathode material	Potential risks
$LiCoO_2$	The degradation of $LiCoO_2$ cathode material will happen if reacting with water, oxidant and acid. Under rigorous reactions, the $LiCoO_2$ cathode material will transform into cobalt or lithium oxides with high toxicity
$LiMn_2O_4$	The thermal decomposition of $LiMn_2O_4$ will generate a lot of oxygen and toxic gas can also generate when $LiMn_2O_4$ interacts with reductant and organic solvents
$LiNiO_2$	The thermal decomposition of $LiNiO_2$ will generate oxygen, nickel oxide and lithium oxide
$LiFePO_4$	The high input energy in industrial processing will lead to environmental burdens

Table 2 The potential risks on environment of lithium salt

Lithium salt	Potential risks
$LiPF_6$	$LiPF_6$ is a highly corrosive chemical. The thermal degradation of $LiPF_6$ will generate PF_5, which is an irritating and toxic gas. In air with high humidity, it will further convert to corrosive acid HF. $LiPF_6$ can be also combustible with the presence of phosphorus oxide
$LiBF_4$	$LiBF_4$ shares the similar physicochemical properties with LiPF6 and it will transfer into HF with the presence of water or acid
$LiClO_4$	$LiClO_4$ is strong oxidant which can easily burn with the presence of oxygen or chlorine gas
$LiAsF_6$	$LiAsF_6$ is incompatible with strong oxidizing agents and can also react with acid, releasing HF
$LiCF_3SO_3$	$LiCF_3SO_3$ can react with acid and oxidants to release HF and is combustible in multiple gases including carbon monoxide, carbon dioxide and sulfur dioxide

2.3 Lithium Salt

The type of lithium salt and their potential risks are listed in Table 2. High corrosive acid like HF can be easily generated if lithium salt containing fluorine ions reacts with water or acid. Lithium salts containing ClO_4^- are highly oxidative and combustible.

2.4 Solvent

The commercialized solvent including DMC, DME, EMC, PC and EC can also be combustible and can react vigorously with carbon monoxide and carbon dioxide. The solvent will decompose and transform into aldehyde or alkone when it is heated. Hence, it may also have potential risks for the batteries operated at high temperature or recycled by pyrometallurgy method.

2.5 *Separator and Binder*

PP and PE are used as separator due to its good electron insulation, ion conduction and easy processing. The potential risks for these separators are its intensive reaction with organic acid, aldehyde and carbon monoxide. The commercialized binder is PVDF in LIBs. It can also decompose and generate HF when heated.

In conclusion, the wide varieties of components of spent LIBs show different potential risks. Therefore, the waste of LIBs should be well detached before recovery and treated separately, and branched recycling process should be designed owing to their different properties.

3 The Evaluation of the Impact of LIBs Waste on Environment

The assessment of the environmental impact of LIBs waste is significant for optimizing the design of recycling process. As for the comprehensive assessment, life cycle thinking is needed, which considers all the environmental impact factors along the whole life cycle of the assessment objects [27]. The specific assessment methodology and evaluation instances are introduced as follows.

3.1 *Impact Assessment Methodology*

Life cycle assessment (LCA) method, a still evolving assessing method, was used to evaluate the possible impact of LIBs product on environment, by considering the harmfulness of the substances to human or nature which exceeds a certain threshold [20, 21]. To obtain the most accurate evaluation results, the input substances and energy in whole lifetime of a product, including the manufacturing, usage and disposal, are quantified and evaluated. Besides, all the elementary flows should be also considered before the implementation of LCA methods, which includes the emission of the potential waste to air, soil and water. However, for some of the emission processes, the release of waste to landfill for example, the harmfulness remains unclear due to their long-time effect. This brings complexity in impact assessment by LCA methods and more factors should be counted into avoid potential inefficiency of LCA method. In this consideration, life cycle impact assessment method (LCIA) was developed which introduces appropriate factors to evaluate the potential impact of substances for long period.

Different calculation methods are used for the implementation of LCIA, which aims to evaluate on various aspects of the product such as energy consumption, ecosystem damage, and abiotic depletion based on specific needs.

Cumulative energy demand (CED) was selected as a representative metric to evaluate the impact of LIBs waste based on environmental considerations, by measuring the direct and indirect energy input throughout the lifetime of LIBs. Aiming to obtain the precise evaluation results, the different energy requirements were well distinguished. For example, there are different requirements for renewable resources and non-renewable ones. The function of CED method is not limited to the impact assessment. It can also serve as a screening indicator for the potential environmental burdens and can also be a reference to compare with the results of LCA only considering primary energy demand. Besides, the errors can be also checked by the implementation of CED method owing to its easiness for judgement.

The Center of Environmental Science of Leiden University proposed a guide to conduct LCA based on the ISO standards, which was well known as CML method thereafter [22]. The ISO standard list includes substances derived from nature or substances released to nature. To fully use this standard to carry out LCA project, several impact factors and characterization methods are introduced in this guidance, to set up a link between the ecoinvent data and impact assessment factors. The impact categories consist of problem-oriented factor and damage-oriented factor. The problem-oriented approach implemented in CML method was further elaborately classified into subcategories, including the acidification potential, climate change, eutrophication potential, freshwater aquatic eco-toxicity, human toxicity, land use, marine aquatic eco-toxicity, photochemical oxidation, resources, stratospheric ozone depletion, terrestrial eco-toxicity, freshwater sediment eco-toxicity, malodours air, marine sediment eco-toxicity and ionizing radiation. There are still some uncertainties of this method despite the elaborate classifications. This includes uncertain factors for some chemicals including hydrogen fluoride and beryllium, in human toxicity, marine sediment eco-toxicity, etc.

Eco-indicator 99 is also an established method for life cycle impact assessment, especially on the substances contributing to the environmental burdens like greenhouse effect and gas pollution. The evaluation results can be optimized by following the three steps: (i) The damage factors of the resources consumption and pollutants are first calculated. (ii) Based on the level of damage categories, the damage factors are further normalized. (iii) Weighting factors are then calculated for the damage factors.

The Global Warming Potential (GWP) was developed to compare the effects of different gases on global warming. Specifically, it is a measure of how much energy will be absorbed by an emission of 1 ton of gases over a given period of time, in comparison to 1 ton of carbon dioxide (CO_2) emissions. High global warming potential value means the gases give lots of heat to the earth during the given period. The time period normally used for global warming potential is 100 years. GWP provides a common unit of measurement that enables analysts to add emission estimates for different gases (for example, to prepare national greenhouse gas inventories) and allows decision makers to compare emission reduction opportunities across sectors.

Abiotic depletion potential (ADP) is a type of evaluation method specifically assessing the potential depletion for abiotic resources [25]. The abiotic resources mainly contain non-living natural resources like fossil energy resources and minerals, the resources regenerated within human lifetimes like groundwater and soil and resources constantly regenerated like wind, river water and solar energy. Owing to the differences of the capability of resources regeneration, the renewability is an important parameter in ADP evaluation. Besides, the decreasing availability should be well quantified in ADP model, which is determined by the remnant reserve and the annual de-accumulation (equal to annual production minus annual regeneration).

Environmental priority strategy (EPS) is an environmental-accounting method to assess the impacts including diversity of biosystem, resources and human health. It is a suitable tool especially for the product designers, to assess the added value of the product and propose the development strategy based on environmental considerations.

The tool for the reduction and assessment of chemical and other environmental impacts (TRACI), containing multiple midpoint impact categories, was also a frequently used method for impact assessment. The ozone depletion, global warming, acidification, eutrophication, photochemical oxidation (smog), ecotoxicity, human health toxicity by air pollutants and human health toxicity by carcinogenics or non-carcinogenics are all included in TRACI method. However, it does not take resource consumption into account. This method was developed specifically for the United States and the input parameters of this method were also based on the locations in the United States.

3.2 Impact Assessment of Spent LIBs

Obviously, the material compositions of various commercialized batteries are totally different, and the impact assessment of different LIBs should be evaluated separately. However, in some cases, the waste of LIBs from recycling industry is a mixed stream, with multiple base metals compositions, which increases the difficulties of recycling. Such barriers can be better overcome by analyzing the different cathode chemistry in LIBs, followed by overall processing, to achieve the best recovering efficiency. A composition analysis of different 18,650 LIBs is shown in Fig. 1, including $LiCoO_2$, $LiMnO_2$, $LiFePO_4$ and mixed metal batteries ($LiMO_2$) [19]. It can be found that base metals still contribute most to the total weight, and among them, steel make up the largest portion of base metals, since most of the packages of LIBs is made of steel. As for the batteries with mixed metal cathode, the commercialized package is made of aluminum. By utilizing the impact methodology of CED, the embodied energies of the components in different 18,650 cells are calculated.

It can be observed from Fig. 2 that the total CED value of $LiFePO_4$ and $LiMn_2O_4$ are smaller than the other two types, about 35% less, showing a less energy consumption and less impact on environmental burden. Besides, aluminum contributes a large portion of CED value, which is due to the large energy consumption in the preparation of aluminum via Hall-Heroult method. It accounts for about 25% of total energy in $LiCoO_2$ cathode and mixed metal cathode, and about 45% of total energy is derived from the aluminum metal in $LiFePO_4$ cathode. Except from aluminum,

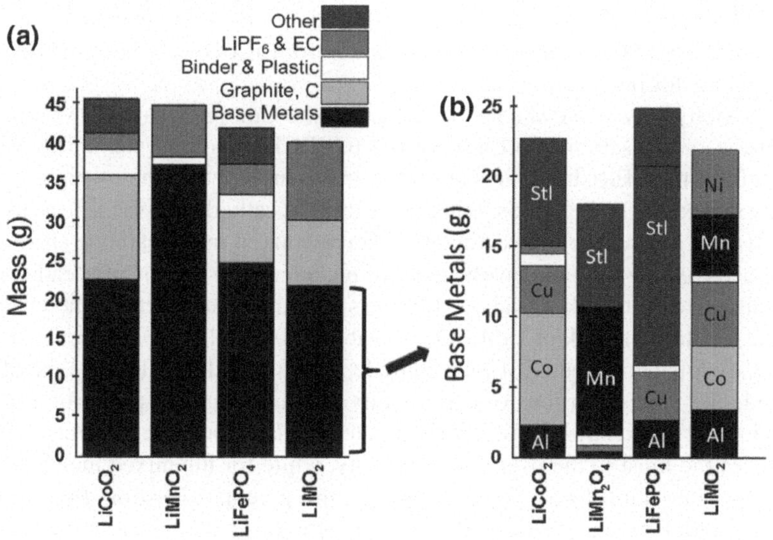

Fig. 1 The compositions of different 18,650 cells by breaking down in total (**a**) and **b** by breaking down for base metals. Reproduced with permission [19], Copyright 2019, Elsevier

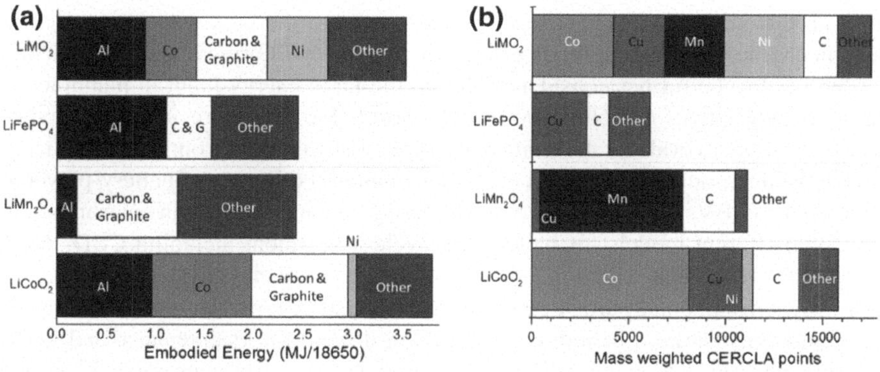

Fig. 2 **a** The embodied energy of the components in different 18,650 cells based on CED calculations. **b** The CERCLA points of different 18,650 cells to evaluate the potential eco-toxicity. Reproduced with permission [19], Copyright 2019, Elsevier

carbon and graphite also contribute a lot to the total CED value, which was about 25% for $LiCoO_2$ cathode and 42% for $LiMn_2O_4$ cathode. The large energy contribution is owing to the consumption of energy of carbon black and graphite in purification step. In addition to these two components, the other contributors of mixed metal cathode are lithium (5%), copper (3%), manganese (6%) and steel (5%).

As for the $LiFePO_4$ cathode, the steel, lithium and copper accounted for total energy about 14, 8 and 5%, respectively. For the $LiMn_2O_4$ base case, manganese accounted for about 22% of total energy, followed by 11% lithium and 8% steel. The main contributor of CED value of other category for $LiCoO_2$ was lithium (10%) and steel accounts for only 5%.

Eco-toxicity was also evaluated by adapting the Comprehensive Environmental Response, Compensation, and Liability Act (CERCLA) point system. The calculation results of the CERCLA points of each component based on their weight in spent LIBs are shown in Fig. 2b. It was found that $LiFePO_4$ cathode shows the least toxicity on ecosystem among the four cathodes, whereas mixed metal cathode showed the highest eco-toxicity with the involvement of nickel. As for the $LiCoO_2$ cathode, the potential eco-toxicity risk was about 10% less than the mixed metal cathode and the potential eco-toxicity risk of $LiMn_2O_4$ was about 70% of $LiCoO_2$ cathode case. As for the individual contributor of each cathode, it was found that the base metals still contribute the largest portions of eco-toxicity risk, although the proportion of their weight in total mass may be low. For the single metal cathode, the base metal always contributes the most to the overall eco-toxicity, while for the mixed metal cathode, different metals almost contribute equally to the potential hazards. Phosphorous, which has a high potential risk on the environment, does not appear in the $LiFePO_4$ case, owing to its existence as oxide in spent LIBs. Still, it should be cautiously treated in the recycling process because it may be reduced to elemental phosphorous and thus be highly toxic to the eco-system.

With the rapid development of electric vehicles industry, large amounts of LIBs were produced, used and disposed. In comparison to the cars with internal combustion engine, whether the mode of electric vehicles will bring less environmental burden need to be further considered, due to the large energy input in manufacturing LIBs and tedious recycling process of spent LIBs. Therefore, systemic studies need to be done to evaluate the contributions of LIBs to the environmental impact in electric vehicle industry. Notter et al. [28] evaluated the impact of battery-powered electric cars (BEVs) and internal combustion engine cars (ICEVs) on environmental burdens. Different models for battery life cycle assessment including CED, ADP, GWP, EI99 H/A were used to evaluate the impact in many ways. The evaluation result of environmental burden caused by BEVs and ICEVs, which takes multiple areas into consideration including road, gilder, drive-train, maintenance and LIBs, was shown in Fig. 3. Generally, BEVs showed a less impact on environmental burden than ICEVs. The environmental burden of ICEV was about 37.47% higher than BEV by using ADP method, about 55.3% higher than BEV by using GWP method, about 23.5% higher than CED method and about 61.6% higher than BEV by using EI99 H/A method. As for the impact of LIBs, the contribution to the overall impact was almost two times higher calculated by EI99 H/A method than calculating by other methods.

The environmental burdens of the main components of LIBs ($LiMn_2O_4$ as cathode) were also assessed by these four methods and the results were shown in Fig. 4. It can be observed that the main environmental burden still came from the anode material, cathode material and battery pack of LIBs no matter what model they used, as indicated in left array in each column. Since copper foil and aluminum foil were used as current collectors and battery pack was made of steel or aluminum in most

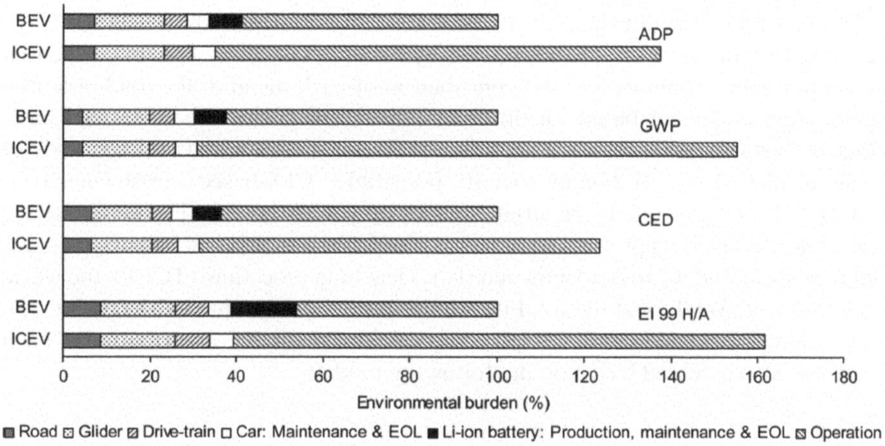

Fig. 3 The environmental burden assessed by CED, ADP, GWP and EI99 H/A methods. Reproduced with permission [28], Copyright 2010, American Chemical Society

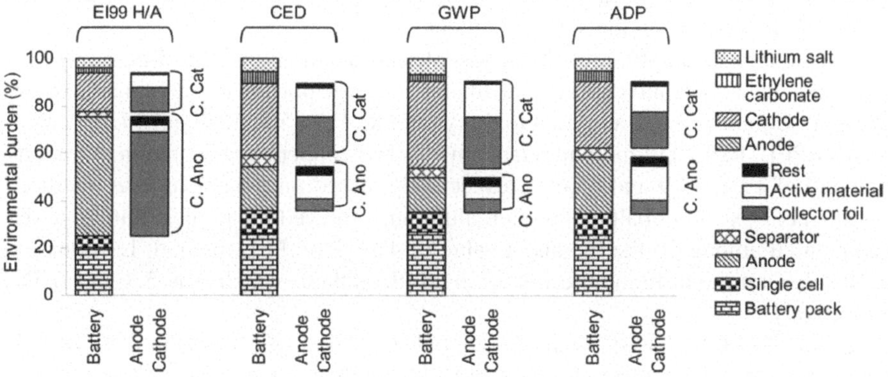

Fig. 4 The environment impact of different components in LIBs assessed by CED, ADP, GWP and EI99 H/A methods. The left array in each column shows the contributions of different components in battery, while the right array shows a closer look of the contribution of each component in electrode materials. Reproduced with permission [28], Copyright 2010, American Chemical Society

cases, the production and recovery of these materials greatly affect the environment. Others including the lithium salt, solvent (EC), separator, single cell play a minimal role. As for the individual components, they contribute differently to the overall impact according to different methods. Copper contributed the largest portion in anode calculated by EI99 H/A method, while for other methods, it contributed much less portion. Graphite in anode showed a larger impact than copper when evaluating by ADP and CED method. As for cathode, aluminum was still the dominant contributor owing to huge energy cost of production, followed by the active material and the rest.

The environmental impact of the LIBs waste, which collected from the facility for recycling cellphones, were also evaluated by Daniel et al. [29]. The total potential impact to the environment and the contribution of each metal to the total potential impact were evaluated based on different impact categories. These impact categories consist of abiotic resource depletion potential (LCIA-based), human toxicity potential (hazard-based), human toxicity potential (LCIA-based), freshwater ecotoxicity (LCIA-based) and terrestrial ecotoxicity (LCIA-based). The study started with chemical leaching process and the waste of LIBs was first leached by standardized tests including the toxicity characteristics leaching procedure (TCLP), the waste extraction test (WET) and the total threshold limit concentration (TTLC) [30, 31]. Subsequently, the total potential impact was calculated, and the contribution of each metal was also assessed based on the following model:

$$P_i = C_i \cdot W \cdot Wf_i$$

P_i represents various types of potential for the metal, which depends on the method used. C_i represents the weight percentage of the total weight of the battery. W represents the total weight of the battery, while Wf_i is the weighting factor of the metal i, which also depends on the selected model.

Based on the calculation results, it was observed that in the hazard-based evaluation, the metal cobalt and copper contributed to a large portion of the human toxicity potential. Cobalt showed a human toxicity potential of 35.3, 50.4 and 17.7% calculated by PEL, REL, TPI method, respectively. The copper showed a human toxicity potential of 19.1, 13.9 and 65.6%, respectively, based on the above three methods in sequence. Except cobalt and nickel, aluminum showed the highest human toxicity potential among all the metals, a value of 42.3% by PEL method. Lithium and nickel also exhibited ignorable impact by TPI method, which was 5.4 and 6.5%, respectively.

As for the LCIA-based evaluation, cobalt had the highest potential for abiotic resource depletion calculated by EPS and TLV method, which was 63.7 and 51.2%, respectively. However, with the implementation of CML method, copper showed the highest value (74.6%) of the potential for abiotic resource depletion. In terms of human toxicity potential, the emission of cobalt metal to air and to soil had high potential toxicity to human, with a contribution of 79.2 and 41.9%, respectively, based on the CML method. Besides, the emission of copper to the air also brings toxicity to human, showing a potential value of 16.1% by CML method. The contribution of metal to human toxicity potential derived from TRACI method were different from CML method, as the emission of nickel to water showed a high toxicity potential of 21.0%. Freshwater ecotoxicity potential and terrestrial potential were also evaluated. Cobalt showed a dominant freshwater and terrestrial toxicity potential, about 79.0 and 92.0%, respectively, based on the CML calculation. However, by TRACI calculation, copper owned the largest contribution to the freshwater ecotoxicity potential (85.4%).

The environmental impacts during the transportation of spent LIBs were also evaluated, since massive energy input is needed for the collection and delivery of spent LIBs to recycling plant. It was concluded that by optimizing geographic location of recycling plants and reducing transportation distances, the environmental impact caused by transportation can be significantly reduced. For example, Anna et al. [32] assessed the environmental impacts of recycling portable LIBs at different locations. It was found that the recycling of batteries within Australia instead of Europe would reduce the emissions of carbon dioxide approximately 300 kg per ton of spent batteries. Besides, GWP increases about 45% compared to the local LIBs recycling.

4 The Impacts of Recycling Process on Environment

Various recycling techniques have been designed to realize the effective recovery of spent LIBs. Typically, a recycling process may contain physical separation, chemical leaching, neutralization, precipitation, filtering and purification. Despite its effectiveness to recover valuable products from spent LIBs with high efficiency, the input of energy, reactants and the output of byproducts may also cause environmental burden, which needs to be systematically evaluated. In this section, we discussed the environmental impacts of production and consumption of leaching agents, of the post treatment of remnant liquor and byproducts generated in recycling process, and of input energy, etc. And the possible approaches to reduce the environmental burden are also discussed.

4.1 The Impact of Added Reactants on Environmental Burden

There are different types of added reactants in different recycling steps. In chemical leaching step, large amounts of leaching agents are needed to ensure the effective leaching. The frequently used leaching agents are acids including sulfuric acid [33], phosphoric acid [34], hydrochloric acid [35], nitric acid [36], chloridric acid [37] assisted with oxidative agents like hydrogen peroxide [38, 39]. When discussing about the environmental impacts of leaching process on environmental burden, the impacts from the production of leaching agents and the treatment of excess of leaching agents should be well considered. As for the acid production, the environmental impact of the production of some of the acids have been evaluated [17, 23, 40, 41]. For example, in the environmental evaluation of toxic elements during the phosphoric production, it was found that the main environmental concern is uranium, which is a highly radiotoxic element, showing the highest toxicity contribution to the environment (23%). Besides, the waste generated in the production of phosphoric also contained elements including Ba, Cu, Sr, Cd, Ni and Zn, showing a potential toxicity contribution of 21, 6, 6, 5, 5, 2%, respectively. The environmental impact of the production of sulfuric acid was also studied by LCA method and the results were shown in Fig. 5. It can be observed that the sulfur treatment step contributed to

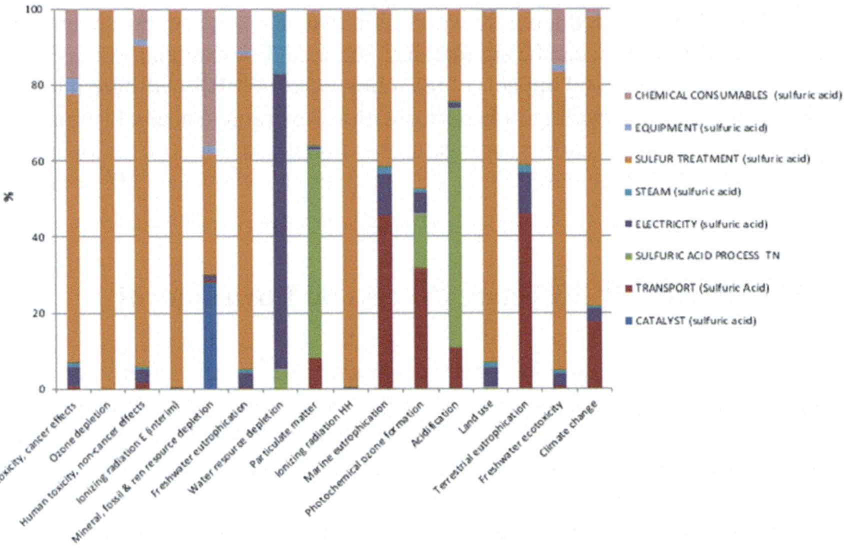

Fig. 5 The evaluation of the environmental impacts of the sulfuric acid production based on different categories. Reproduced with permission [17], Copyright 2019, Springer Nature

almost every category of environmental impacts and creates a lot of environmental burden, especially on the human toxicity, ozone depletion, ionizing radiation, freshwater eutrophication, and land use. This is due to the huge energy consumption in petroleum desulfurization. In terms of the environmental impact of remnant leaching agents, it is hard to evaluate from a generic level since it depends on the different recycling techniques. Moreover, lots of leaching agents will transfer into metal salt and be collected at the end. Such treatment also brings complexity for the evaluation and advanced modelling method is needed to consider all the impact factors to avoid discrepancy of the environmental evaluation results.

Apart from leaching agents, the use of other added reactants in the recycling process including acid or alkaline for neutralization, metal salt for precipitation and organic solvent for solvent extraction also lead to environmental burden. Most of their impacts have not been systematically studied yet, therefore future research is needed to evaluate their environmental impacts and optimize the recycling technique.

4.2 The Impact of Input Energy on Environmental Burden

The input energy is also an important parameter for evaluating potential environmental burden. Li et al. [42] evaluated the energy consumption in a typical recycling process. Purchase energy intensity data, which is the intensity of the energy consumption of fuel at recycling plant, was incorporated to evaluate the environmental

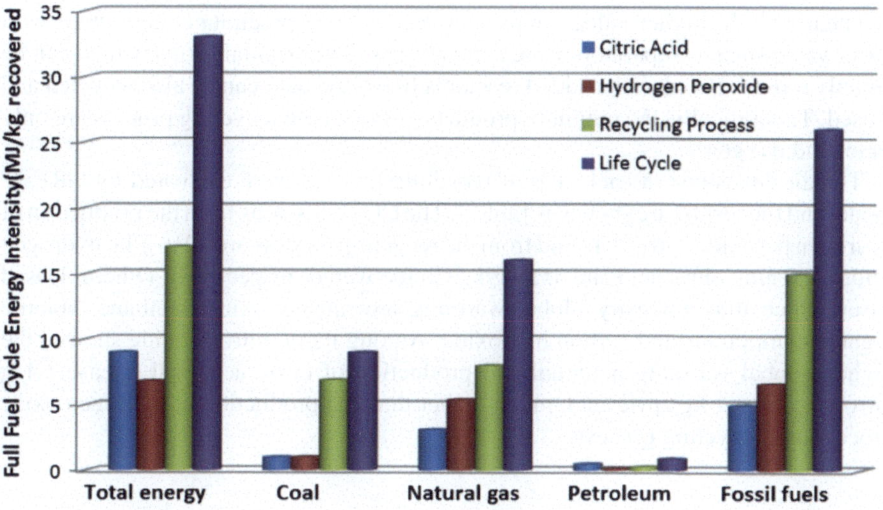

Fig. 6 FFC energy density for citric acid, hydrogen peroxide and the recycling process. Reproduced with permission [42], Copyright 2019, Elsevier

burden of each recycling step. It was found that in the cobalt recovery, the calcination step contributed most to the purchased energy density among all the steps, followed by crushing and grinding, whereas steps like discharging, NMP soak and leaching had minimal purchased energy intensity. By employing Argonne national laboratory's greenhouse gases, regulated emissions, and energy use in transportation (GREET) model, full cell cycle (FFC) energy intensity of recycling process was calculated, and the results were shown in Fig. 6. The citric acid and hydrogen peroxide were introduced in the leaching process and it can be observed that the total energy consumption is about 33 MJ kg^{-1}, including upstreams steps such as coal mining, natural gas, petroleum and fossil fuels collection. Although citric acid had a significant impact on the energy consumption based on the FFC results, most of the citric acid (about 90%) can be recovered in the recycling process and no severe environmental burdens will be caused since it is eco-friendly substance.

4.3 The Impact of Output Byproducts on Environmental Burden

The types of the output byproducts could be various including waste liquor and released gases, and it will cause environmental burden like greenhouse effect. One way to reduce the environmental burden is to decrease the amounts of toxic byproducts in the recycling process. Chen et al. [1] reported the sustainable recovery of metals from spent LIBs and the waste stream was also analyzed in detail. It was found that nearly all the higher-value scraps were recovered as precipitates and scraps with lower value such as separators were recovered as their orignial forms, which can be directly reused. Besides, the added reactants like citric acid can be also recycled and reused. Theoretically, the output byproducts of the whole recycling process are only water and oxygen.

The air emissions in each step of recycling process were evaluated by GREET model and the results are shown in Table 3. The CO_2 emissions from the production of hydrogen peroxide, citric acid and from the recycling process are 1210 g/kg hydrogen acid, 1380 g/kg citric acid and 4559 g/kg Co recovered, respectively. Other released gases which may intensify global warming are nitrous oxide, methane, volatile organic compounds and carbon monoxide. Among them, nitrous oxide showed the highest global warming potential and production of citric acid will release 1.1 g nitrous oxide per kg citric acid, much higher than the production of hydrogen oxide process and recycling process.

Table 3 Full fuel cycle air emissions from production of hydrogen peroxide, citric acid, and from the recycling process. Reproduced with permission [42], Copyright 2013, Elsevier

Pollutant	Emissions from H_2O_2 production (g/kg H_2O_2)	Emissions from citric acid production (g/kg citric acid)	Emissions from recycling process (g/kg Co recovered)
Volatile organic compounds	0.13	1.32	0.87
Carbon monoxide	0.31	1.11	0.80
Nitrogen oxides	1.2	2.9	2.9
Particulate matter (<10 μm)	0.66	1.3	2.1
Particulate matter (<2.5 μm)	0.28	0.46	0.92
Sulfur oxides	1.4	1.9	3.9
Methane	6.9	6.3	8.6
Nitrous oxide	0.01	1.1	0.32
Carbon dioxide	1.210	1.380	4.559
Greenhouse gases[a]	1.386	1.885	4.873

[a]Greenhouse gases include CO_2, CH_4, and N_2O with different global warming potentials of 1, 25 and 298, respectively. And the effect of CO and carbon volatile organic compounds are also included

5 Approaches for Reducing Environmental Burden: Toward Designing Environmental Benign Recycling Process

5.1 The Optimization of Leaching Agent Selection

The production of commonly used leaching acid normally contaminates the environment, which greatly increases the environmental burden of leaching process. Therefore, more environmentally friendly and readily available leaching acid will be a better selection. DL-malic acid ($C_4H_5O_6$) was selected as an effective leaching agent, which owns the following merits.

Firstly, it is a natural organic acid which can be easily isolated from the apple juice. The easy availability of DL-malic acid will save the energy input for industrialized production and decrease the environmental pollution. Secondly, in comparison to commonly used acid, it is easy to degrade under aerobic and anaerobic conditions. The remaining DL-malic acid after chemical leaching can be easily handled. By using the DL-malic acid as leaching agent, nearly 100% lithium and more than 90% cobalt can be extracted from spent LIBs [43]. In another study, mechanochemical approach was used in the recovery of cobalt and lithium [27]. It was an environmental benign process without using strong acid and oxidants in the leaching process. Instead, ethylenediaminetetraacetic acid (EDTA), a kind of organic compound, was used as additive in the co-grinding process, followed by a water leaching process to recover the cobalt and lithium from the spent LIBs effectively. It was demonstrated that 98% of cobalt and 99% of lithium can be recovered. The mechanism studies further demonstrated its effectiveness in leaching process, which is due to the fact that the functional groups atoms of EDTA, including oxygen and nitrogen atoms, can occupy the empty orbit of cobalt and lithium by solid interface reaction, forming a stable and also water-soluble compound.

5.2 The Optimization by Combining Different Methods

Apart from optimizing the selection of leaching agent, the combination of physical approach with chemical exfoliation is also effective to lighten the environmental burden, since the leaching efficiency can be much improved, and the dosage of leaching acid can be also reduced. For example, it was found that the leaching efficiency could be greatly improved by employing ultrasonication in chemical leaching process [44]. This is due to the faster reaction between acid and metal ions under ultrasonication and the particle size of residue became smaller after ultrasonic treatment. As a result, less amounts of leaching agents were used and a higher leaching efficiency was obtained, with a recovery of 96.13% cobalt and 98.4% lithium. In addition, supercritical carbon dioxide extraction was also an effective method to be used in metal leaching [45].

The merits of employing this technique included a faster reaction kinetic and reduce the amounts of reactant. As a result, 98% extraction of cobalt was achieved only in a short reaction time of 5 min and with a small concentration of 4% (v/v) H_2O_2 compared to conventional technique, which required a reaction time of 60 min and a concentration of 8% (v/v) H_2O_2 to reach the same effect.

5.3 The Optimization of Energy Input by Considering Regional Factors

Regional energy structure is also an important factor that needs to be considered for reducing environmental burdens of recycling industry in a specific area. As we discussed above, large energy input are needed to dismantle or crush the spent LIBs and the generation of the consumed energy, especially the electric power, varies greatly from region to region. As for the region in which the generation of the electric power highly relies on the combustion of non-renewable energy like coal and natural gas, the massive use of electric power in recycling process may counteract the environmental benefits that recycling itself brings, since the collection of the non-renewable energy also consumes energy and human resources and the byproducts in the generation of electric power also cause environmental burden. Instead, pyrometallurgical process may be a better option to recover the valuable product by calcination with the improvement of energy efficiency. For the region where electric power generation relies on renewable energy like wind and tidal energy, the electric power is recommended to be used in recycling process because it will not lead to additional environmental burdens. Therefore, by optimizing the types of energy input in recycling process based on regional factors, the environmental burden can be reduced.

6 Conclusion and Perspectives

Great efforts have been made in developing advanced recycling techniques of spent LIBs. High value products could be successfully recycled from spent LIBs with high recovery efficiency. Nonetheless, with less focus on the environmental impact of recycling process, questions on the potential environmental burden of recycling process are raised. In this chapter, we give a systematic overview of this issue. We first list the potential risks of spent LIBs, concluding that the waste material of spent LIBs would better be separated before recovery owing to their various types. Moreover, branched recycling process should be rationally designed due to their different properties. In addition, impact assessment methodology, i.e. life cycle impact assessment method, is introduced to evaluate the potential impacts of LIBs waste in a whole life cycle. The subcategories of this method are also introduced, such as the cumulative energy input/output and the usage of resources, which can evaluate and optimize the proposed recycling technique from a specific viewpoint. The environmental impacts of the recycling process are also discussed, concluding that the added reactants, the input energy and the output byproducts in recycling process can all lead to environmental burden. In this consideration, promising approaches to reduce environmental burdens are proposed, including the optimization of the selection of leaching agent, the combination of different methods to improve efficiency and the optimization of energy input.

More efforts are still required on the optimization of recycling processes from an environmental point of view. First, all the unnecessary environmental burdens creators should be strictly restricted. Next, a proposed recycling technique should be thoroughly evaluated by LCIA method and all the potential impact factors should be taken into considerations, such as the cumulative energy, human toxicity, resource depletion potential, etc. At last, distributions of local resources and energy structures should also be well adjusted to avoid additional environmental burdens.

Acknowledgements This work was fully supported by a grant from the Research Grants Council of the Hong Kong Special Administrative Region, China (Project No. 15222018).

References

1. X. Chen, C. Luo, J. Zhang, J. Kong, T. Zhou, Sustainable recovery of metals from spent lithium-ion batteries: a green process. ACS Sustain. Chem. Eng. **3**, 3104–3113 (2015). https://doi.org/10.1021/acssuschemeng.5b01000
2. X. Zeng, J. Li, L. Liu, Solving spent lithium-ion battery problems in China: opportunities and challenges. Renew. Sustain. Energy Rev. **52**, 1759–1767 (2015). https://doi.org/10.1016/j.rser.2015.08.014
3. A. Chagnes, B. Pospiech, A brief review on hydrometallurgical technologies for recycling spent lithium-ion batteries. J. Chem. Technol. Biotechnol. **88**, 1191–1199 (2013). https://doi.org/10.1002/jctb.4053
4. D. Lisbona, T. Snee, A review of hazards associated with primary lithium and lithium-ion batteries. Process Saf. Environ. Prot. **89**, 434–442 (2011). https://doi.org/10.1016/j.psep.2011.06.022
5. B. Swain, J. Jeong, J. chun Lee, G.H. Lee, J.S. Sohn, Hydrometallurgical process for recovery of cobalt from waste cathodic active material generated during manufacturing of lithium ion batteries. J. Power Sources. **167**, 536–544 (2007). https://doi.org/10.1016/j.jpowsour.2007.02.046
6. L. Li, R.J. Chen, X.X. Zhang, F. Wu, J. Ge, M. Xie, Preparation and electrochemical properties of re-synthesized LiCoO2 from spent lithium-ion batteries. Chinese Sci. Bull. **57**, 4188–4194 (2012). https://doi.org/10.1007/s11434-012-5200-5
7. D.S. Kim, J.S. Sohn, C.K. Lee, J.H. Lee, K.S. Han, Y.I. Lee, Simultaneous separation and renovation of lithium cobalt oxide from the cathode of spent lithium ion rechargeable batteries. J. Power Sources. **132**, 145–149 (2004). https://doi.org/10.1016/j.jpowsour.2003.09.046
8. M. Lu, H. Zhang, B. Wang, X. Zheng, C. Dai, The re-synthesis of LiCoO2 from spent lithium ion batteries separated by vacuum-assisted heat-treating method. Int. J. Electrochem. Sci. **8**, 8201–8209 (2013)

9. H. Nie, L. Xu, D. Song, J. Song, X. Shi, X. Wang, L. Zhang, Z. Yuan, $LiCoO_2$: Recycling from spent batteries and regeneration with solid state synthesis. Green Chem. **17**, 1276–1280 (2015). https://doi.org/10.1039/c4gc01951b
10. D. Bian, Y. Sun, S. Li, Y. Tian, Z. Yang, X. Fan, W. Zhang, A novel process to recycle spent LiFePO4for synthesizing $LiFePO_4$/C hierarchical microflowers. Electrochim. Acta **190**, 134–140 (2016). https://doi.org/10.1016/j.electacta.2015.12.114
11. X.M. He, J.G. Li, Q. Zhang, Preparation of LiNi1/3Co1/3Mn1/3O2 cathode materials from spent Li-ion batteries, s897–s901 (2007)
12. D. Song, X. Wang, E. Zhou, P. Hou, F. Guo, L. Zhang, Recovery and heat treatment of the Li(Ni1/3Co1/3Mn1/3)O_2 cathode scrap material for lithium ion battery. J. Power Sources **232**, 348–352 (2013). https://doi.org/10.1016/j.jpowsour.2012.10.072
13. L. Yao, H. Yao, G. Xi, Y. Feng, Recycling and synthesis of LiNi1/3Co1/3Mn1/3O2from waste lithium ion batteries using d, l-malic acid. RSC Adv. **6**, 17947–17954 (2016). https://doi.org/10.1039/c5ra25079j
14. Y. Weng, S. Xu, G. Huang, C. Jiang, Synthesis and performance of Li[(Ni1/3Co1/3Mn1/3)1-xMgx] O_2 prepared from spent lithium ion batteries. J. Hazard. Mater. **246–247**, 163–172 (2013). https://doi.org/10.1016/j.jhazmat.2012.12.028
15. M.J. Ganter, B.J. Landi, C.W. Babbitt, A. Anctil, G. Gaustad, Cathode refunctionalization as a lithium ion battery recycling alternative. J. Power Sources **256**, 274–280 (2014). https://doi.org/10.1016/j.jpowsour.2014.01.078
16. A.M. Bernardes, D.C.R. Espinosa, J.A.S. Tenório, Recycling of batteries: a review of current processes and technologies. J. Power Sources **130**, 291–298 (2004). https://doi.org/10.1016/j.jpowsour.2003.12.026
17. M. Mami, M. Jeday, N. Hajjaji, Life cycle assessment of sulfuric acid production system in Tunisia. 483–485 (2018)
18. J. Dewulf, G. Van der Vorst, K. Denturck, H. Van Langenhove, W. Ghyoot, J. Tytgat, K. Vandeputte, Recycling rechargeable lithium ion batteries: Critical analysis of natural resource savings. Resour. Conserv. Recycl. **54**, 229–234 (2010). https://doi.org/10.1016/j.resconrec.2009.08.004
19. X. Wang, G. Gaustad, C.W. Babbitt, C. Bailey, M.J. Ganter, B.J. Landi, Economic and environmental characterization of an evolving Li-ion battery waste stream. J. Environ. Manage. **135**, 126–134 (2014). https://doi.org/10.1016/j.jenvman.2014.01.021
20. R. Heijungs, J.B. Guinee, G. Huppes, R.M. Lankreijer, D.H. Udo, A.W. Sleeswijk, A.M.M. Ansems, P.G. Eggels, R. Duit, H.P. Goede, Environmental life cycle assessment of products, **1**: Guide (1992)
21. R. Frischknecht, N.J. Editors, H. Althaus, C. Bauer, G. Doka, R. Dones, R. Hischier, S. Hellweg, T. Köllner, Y. Loerincik, M. Margni, Implementation of life cycle impact assessment methods. Am. Midl. Nat. **150**, 1–151 (2007). http://www.ecoinvent.org/fileadmin/documents/en/03_LCIA-Implementation.pdf
22. E. Johnson, Handbook on life cycle assessment operational guide to the ISO standards. Environ. Impact Assess. Rev. **23**, 129–130 (2003). https://doi.org/10.1016/S0195-9255(02)00101-4
23. S. Belboom, C. Szöcs, A. Léonard, Environmental impacts of phosphoric acid production using di-hemihydrate process: a Belgian case study. J. Clean. Prod. **108**, 978–986 (2015). https://doi.org/10.1016/j.jclepro.2015.06.141
24. J.L. Sullivan, L. Gaines, Status of life cycle inventories for batteries. Energy Convers. Manag. **58**, 134–148 (2012). https://doi.org/10.1016/j.enconman.2012.01.001
25. J.B. Guinée, R. Heijungs, A proposal for the definition of resource equivalency factors for use in product life-cycle assessment. Environ. Toxicol. Chem. **14**, 917–925 (1995). https://doi.org/10.1002/etc.5620140525
26. B. Li, X. Gao, J. Li, C. Yuan, Life cycle environmental impact of high-capacity lithium ion battery with silicon nanowires anode for electric vehicles. Environ. Sci. Technol. **48**, 3047–3055 (2014). https://doi.org/10.1021/es4037786
27. M.M. Wang, C.C. Zhang, F.S. Zhang, An environmental benign process for cobalt and lithium recovery from spent lithium-ion batteries by mechanochemical approach. Waste Manage. **51**, 239–244 (2016). https://doi.org/10.1016/j.wasman.2016.03.006

28. D.A. Notter, M. Gauch, R. Widmer, W.A. Patrick, A. Stamp, R. Zah, R.G. Althaus, Contribution of Li-ion batteries to the environmental impact of electric vehicles, **44**, 6550–6556 (2010)
29. D.H.P. Kang, M. Chen, O.A. Ogunseitan, Potential environmental and human health impacts of rechargeable lithium batteries in electronic waste. Environ. Sci. Technol. **47**, 5495–5503 (2013). https://doi.org/10.1021/es400614y
30. C.D.W.E.T. (WET) P. State, 2005 of California: Sacramento, CA, Waste Classification, (1989)
31. U.E.M. 1311, Toxicity characteristic leaching procedure. 1–4 (1992). https://doi.org/10.15713/ins.mmj.3
32. A. Boyden, V. Kie Soo, M. Doolan, The environmental impacts of recycling portable lithium-ion batteries. Procedia CIRP. **48**, 188–193 (2016). https://doi.org/10.1016/j.procir.2016.03.100
33. X. Chen, Y. Chen, T. Zhou, D. Liu, H. Hu, S. Fan, Hydrometallurgical recovery of metal values from sulfuric acid leaching liquor of spent lithium-ion batteries. Waste Manage. **38**, 349–356 (2015). https://doi.org/10.1016/j.wasman.2014.12.023
34. X. Chen, H. Ma, C. Luo, T. Zhou, Recovery of valuable metals from waste cathode materials of spent lithium-ion batteries using mild phosphoric acid. J. Hazard. Mater. **326**, 77–86 (2017). https://doi.org/10.1016/j.jhazmat.2016.12.021
35. M.A.H. Shuva, A. Kurny, Hydrometallurgical recovery of value metals from spent lithium ion batteries. Am. J. Mater. Eng. Technol. **1**, 8–12 (2013). https://doi.org/10.12691/materials-1-1-2
36. J. Nan, D. Han, X. Zuo, Recovery of metal values from spent lithium-ion batteries with chemical deposition and solvent extraction. J. Power Sources **152**, 278–284 (2005). https://doi.org/10.1016/j.jpowsour.2005.03.134
37. G. Granata, E. Moscardini, F. Pagnanelli, F. Trabucco, L. Toro, Product recovery from Li-ion battery wastes coming from an industrial pre-treatment plant: lab scale tests and process simulations. J. Power Sources **206**, 393–401 (2012). https://doi.org/10.1016/j.jpowsour.2012.01.115
38. S.G. Zhu, W.Z. He, G.M. Li, X. Zhou, X.J. Zhang, J.W. Huang, Recovery of Co and Li from spent lithium-ion batteries by combination method of acid leaching and chemical precipitation. Trans. Nonferrous Met. Soc. China (English Ed). **22**, 2274–2281 (2012). https://doi.org/10.1016/s1003-6326(11)61460-x
39. Y. Huang, G. Han, J. Liu, W. Chai, W. Wang, S. Yang, S. Su, A stepwise recovery of metals from hybrid cathodes of spent Li-ion batteries with leaching-flotation-precipitation process. J. Power Sources **325**, 555–564 (2016). https://doi.org/10.1016/j.jpowsour.2016.06.072
40. Z. Kirova-Yordanova, Application of the exergy method to the environmental impact estimation: the nitric acid production as a case study. Energy. **36**, 3733–3744 (2011). https://doi.org/10.1016/j.energy.2010.12.039
41. R. Pérez-López, A.M. Álvarez-Valero, J.M. Nieto, Changes in mobility of toxic elements during the production of phosphoric acid in the fertilizer industry of Huelva (SW Spain) and environmental impact of phosphogypsum wastes. J. Hazard. Mater. **148**, 745–750 (2007). https://doi.org/10.1016/j.jhazmat.2007.06.068
42. L. Li, J.B. Dunn, X.X. Zhang, L. Gaines, R.J. Chen, F. Wu, K. Amine, Recovery of metals from spent lithium-ion batteries with organic acids as leaching reagents and environmental assessment. J. Power Sources **233**, 180–189 (2013). https://doi.org/10.1016/j.jpowsour.2012.12.089
43. L. Li, J. Ge, R. Chen, F. Wu, S. Chen, X. Zhang, Environmental friendly leaching reagent for cobalt and lithium recovery from spent lithium-ion batteries. Waste Manage. **30**, 2615–2621 (2010). https://doi.org/10.1016/j.wasman.2010.08.008
44. L. Li, L. Zhai, X. Zhang, J. Lu, R. Chen, F. Wu, K. Amine, Recovery of valuable metals from spent lithium-ion batteries by ultrasonic-assisted leaching process. J. Power Sources **262**, 380–385 (2014). https://doi.org/10.1016/j.jpowsour.2014.04.013
45. D.A. Bertuol, C.M. Machado, M.L. Silva, C.O. Calgaro, G.L. Dotto, E.H. Tanabe, Recovery of cobalt from spent lithium-ion batteries using supercritical carbon dioxide extraction. Waste Manage. **51**, 245–251 (2016). https://doi.org/10.1016/j.wasman.2016.03.009

The manufacturer's authorised representative in the EU is Springer Nature Customer Service Centre GmbH, Europaplatz 3, 69115 Heidelberg, Germany. If you have any concerns regarding our products, please contact ProductSafety@springernature.com

Printed and bound by CPI Group (UK) Ltd, Croydon, CR0 4YY
05/10/2026
02228145-0003